Mechanothermodynamics

Leonid Sosnovskiy · Sergei Sherbakov

Mechanothermodynamics

Springer

Leonid Sosnovskiy
S&P Group TRIBOFATIGUE Ltd.
Gomel
Belarus

Sergei Sherbakov
Belarusian State University
Minsk
Belarus

Translated by E.A. Zharkova

ISBN 978-3-319-24979-7 ISBN 978-3-319-24981-0 (eBook)
DOI 10.1007/978-3-319-24981-0

Library of Congress Control Number: 2015954581

Springer Cham Heidelberg New York Dordrecht London

Printed on acid-free paper

Springer International Publishing AG Switzerland is part of Springer Science+Business Media
(www.springer.com)

New knowledge elucidates space and time

L. Sosnovskiy

In front of you is a small book. It shows the birth of a new physical discipline—*mechanothermodynamics*. This happened when *two small bridges* were built. The first one is the *Tribo-Fatigue entropy* which paved the way from thermodynamics to mechanics. The second one is a fundamental understanding of *irreversible damageability of all things* that paved the way from mechanics to thermodynamics. This path and this way mutually ran through *dielectical* Λ*-interactions* between damages due to loads of different nature (mechanical, thermodynamical, electro-chemical, etc.) and characteristic entropy components (thermodynamical, Tribo-Fatigue, chemical, etc.). A mechanothermodynamic system, as a typical and important *component of the real world*, and its evolution thus become objects of study in *natural science* and requires a *philosophical understanding*.

Contents

Preface

This book presents the formulation of the fundamentals of mechanothermodynamics and analyzes its four principles. The first principle establishes the generalized law of damageability, the second establishes its main cause; the third, its scale; and the fourth characterizes the interrelation of motion, damage, and information. It is shown that in specific (living) systems the information accumulation in time leads to the emergence of elements of intelligence. A generalized model of energy and entropy states of a mechanothermodynamic medium, which in general is a continuum (liquid, gas) with distributed deformable solids and, hence, with damageable solids, is developed. The generalized theory of A-evolution of systems (by damageability) is outlined and further research directions are analyzed.

Chapter 1
From Theoretical Mechanics to Mechanothermodynamics

Any scientific discipline serves and aims at understanding and describing these or those regularities and features of certain phenomena, situations, events caused by the existence of some real or thinkable objects that reveal specific properties [5].

Based on the considerations that the study of a new object, as a rule, generates a new scientific discipline as applied to mechanics, the hierarchy of objects can be built. Figure 1.1 illustrates the very simplified scheme (suffice it to say that liquid and gas continua, etc., are absent). Using Fig. 1.1, one can determine the place of mechanothermodynamics as a new branch of knowledge [5].

When a material object was mentally presented as a dimensionless and structureless *point* able only to move in space and time along any trajectory and in any direction, the birth of *theoretical mechanics* was required to understand and describe the whole variety of motions of such a physically unreal object. The notion of the "point scale" turned theoretical mechanics into a useful science: it became possible to make a correct analysis of motion, for example, of points—planets or points—electrons, i.e., of huge objects of the Universe and unimaginably small objects of the microcosm. If "big points" have mass, then it goes so far as to establish, for example, the interaction laws of celestial bodies in the process of their motion, etc. Space flight mechanics, mechanics of mechanisms, and machines—everything that moves is analyzed with the use of theoretical mechanics methods.

A set of points interconnected in a certain way is a continuum; one of its particular types is a *solid body*. It possesses special (specific) properties: rigidity and strength. When it has been found that under the action of various external loads the points of a solid are able to move or shift relative to each other, the idea of a new object—*deformable solid body* has appeared. Hereinafter it will be called a solid. Naturally, there was a need to create *deformable solid mechanics* in an effort to learn how the solid stress–strain state at any point should be examined and, finally, to understand and describe the regularities and features of size and distortion changes of the solid as a whole. The deformable solid was called simply the *material* or the *sample* or the *structural element* depending on the specific objectives of research. Specific properties of these objects are studied within the framework of such disciplines as *mechanics of materials, structures, composites, soils*, etc., *damage and failure mechanics* (at static, impact, cyclic, etc. loadings), *mesomechanics, micromechanics*, etc. In addition, the regularities, features, and

L. Sosnovskiy and S. Sherbakov, *Mechanothermodynamics*,
DOI 10.1007/978-3-319-24981-0_1

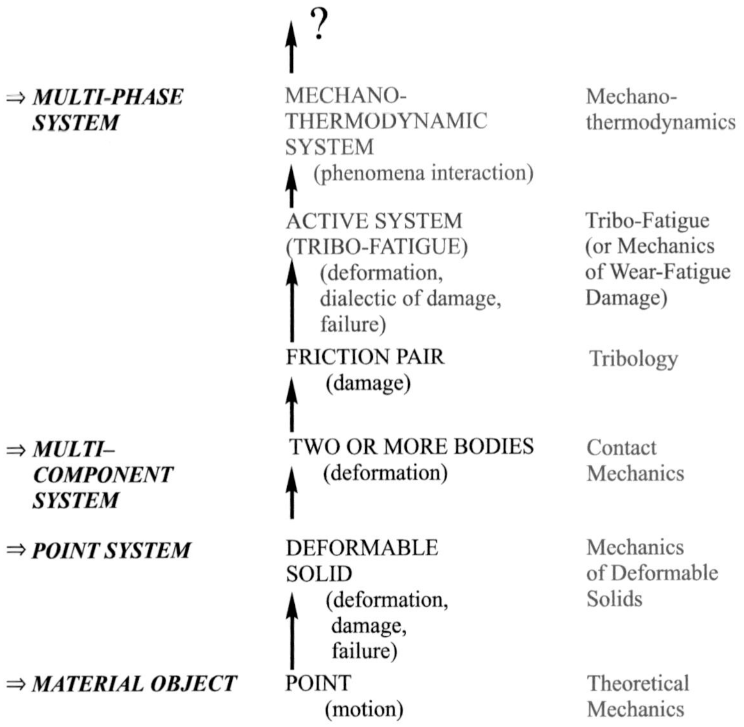

Fig. 1.1 Simplified hierarchical structure of some objects studied in mechanics: from simple to complex

consequences of reversible (*theory of elasticity*) and irreversible (*theory of plasticity*) motions of points of deformable solids are learned. The latter are also provided with a diversity of specific properties, for example, viscoelasticity, elasto-viscoplasticity, etc. Thus, *deformable solid mechanics* has become one of the most powerful tools to study the behavior of real objects in different operating or testing conditions.

A deformable solid is only one of the components of numerous and various *mechanical systems*. Even the simplest case of *compression of two motionless solids* caused the development of a new approach in the theory of elasticity and it was called the *contact problem*. The theory was the beginning of *mechanics of contact interaction of solids* (components) at static, impact, cyclic, and other loads. The next object is the *friction pair*, whose main characteristic is a *relative motion of two solids* at contact load. A special scientific discipline—*tribiology,* the main task of which is to study the regularities and features of friction and surface damage of different materials at sliding, rolling, slippage, impact, etc. Basically, any friction pair is a *multicomponent system*: the *third solid* is inevitably created. It is formed in the moving contact region due to a lubricant and/or tribo-destruction products of thin surface layers of contacting bodies.

More complex than the friction pair is the peculiar object—the *active system* (Tribo-Fatigue) [3], whose concept has been introduced recently (in the late twentieth century). It is the name of any mechanical system that *supports and transits a working cyclic load and in which the friction process occurs simultaneously in any of its manifestations—at sliding, rolling, impact,* etc [2, 4]. In other words, the active system is a friction pair, one element of which is subjected to volumetric re-variable deformation. Such systems are characteristic for *complex wear-fatigue damage*. They are based on the kinetic interaction of the phenomena of fatigue, friction, wear, erosion, corrosion, etc. Naturally, the detection of a new specific object has led to the scientific discipline. Its short name is *Tribo-Fatigue* (from the Greek word "tribo"—*friction*, from the French word "fatigue"—*fatigue*) [3]. Its longer name is *mechanics of wear-fatigue damage* [5] or *mechanics of Tribo-Fatigue systems* [1].

Consider the basic example of stress-strain state in an arbitrary point $M(x, y, z)$ of an active system (Sosnovskiy–Zhuravkov–Sherbakov model) [1, 2]:

$$\sigma_{ij} = \sigma_{ij}^{(n)} + \sigma_{ij}^{(\tau)} + \sigma_{ij}^{(b)}, \quad i, j = x, y, z, \tag{1.1}$$

where $\sigma_{ij}^{(n)}, \sigma_{ij}^{(\tau)}, \sigma_{ij}^{(b)}$ are the stresses caused by the normal and tangential contact tractions, and the non-contact loads respectively.

Figure 1.2 shows stress states in the shaft for the action of elliptically distributed normal contact tractions $p(x, y) = p_0\sqrt{1 - x^2/a^2 - y^2/b^2}$, tangential (frictional) tractions along x axis $q(x, y) = fp(x, y)$ (p_0 is the maximum pressure in the center of the contact area, a and b are the dimensions of the semiaxes of the elliptic contact area) and non-contact bending. There is the significant change of distributions and extreme values (up to 50 % comparing to $\sigma_{xx}^{(n)}, \sigma_{int}^{(n)}$ for pure normal contact) of σ_{xx} stresses and stress intensity in the volumetric neighborhood of the contact region due to the action of friction force($\sigma_{xx}^{(n+\|a)}, \sigma_{int}^{(n+\|a)}$) and non-contact bending ($\sigma_{xx}^{(n+\|a+b)}, \sigma_{int}^{(n+\|a+b)}$) in the active system.

Figure 1.1 uses successive arrows for illustration of the complication of objects studied by mechanics. The next new object was represented by a multiple phase system—a *mechanothermodynamic system*. To study it, the methods of mechanics, as well as of thermodynamics are not enough.

The fundamentals of mechanothermodynamics are given in this book and four of its principles are formulated: the first principle establishes the generalized law of damageability, the second—its main cause, the third—its scale and the fourth characterizes the interrelation of motion, damage, and information. The analysis is made according to the energy representations (in mechanics and Tribo-Fatigue) and is based on the concept of entropy (in thermodynamics and Tribo-Fatigue). This has made it possible to reveal and investigate new regularities of the behavior and evolution of the mechanothermodynamic system.

A generalized model of energy and entropy states of a mechanothermodynamic medium, which generally is a continuum (liquid, gas) with distributed deformable

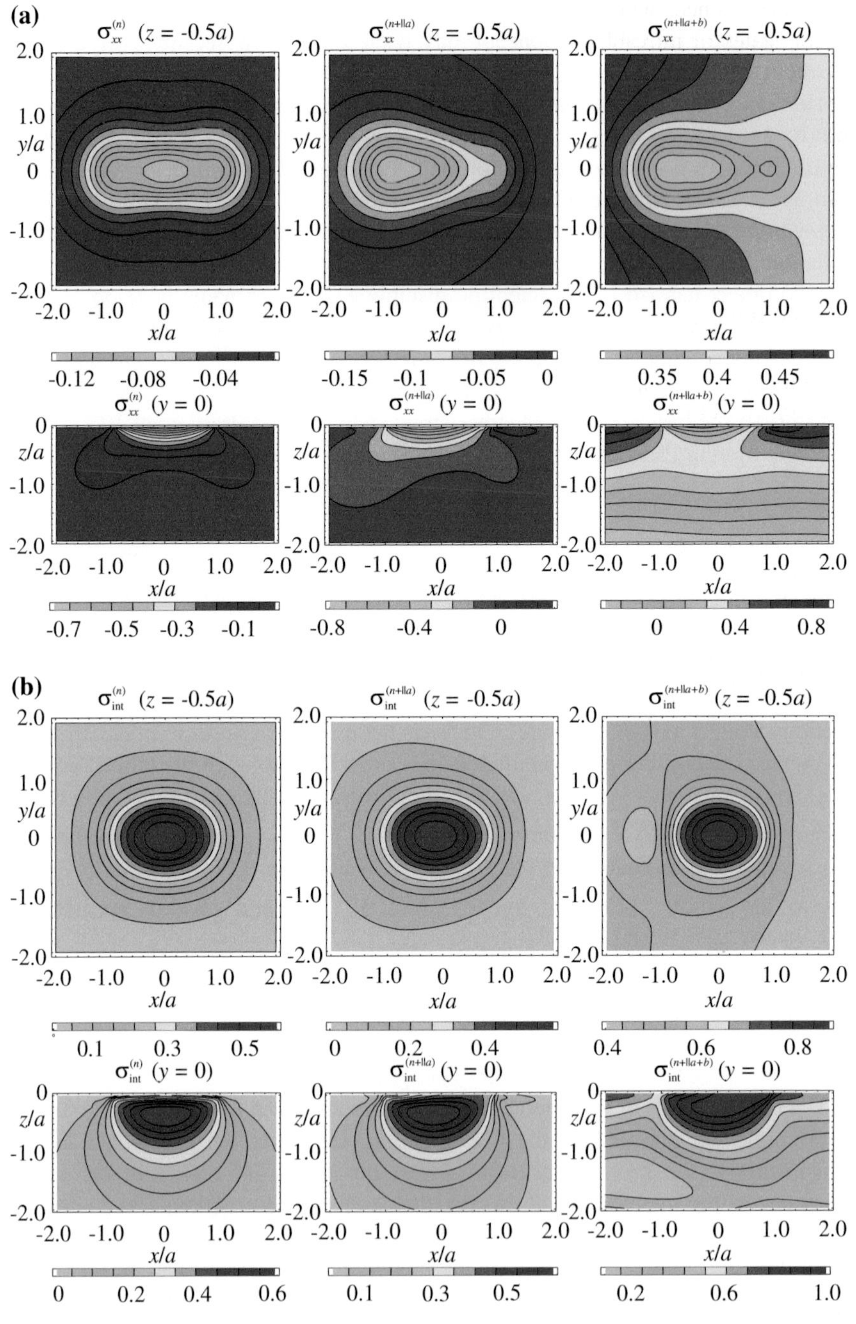

Fig. 1.2 Comparison of fields of σ_{xx} stresses **a** and σ_{int} stress intensity **b** in Tribo-Fatigue system for different loads (L.A. Sosnovskiy, S.S. Sherbakov)

solids and, hence, with damageable solids, is developed. The problem on information states of movable and damageable system is stated and solved in the first approximation. Finally, some directions for further research are analyzed.

The authors are grateful to the reviewers M.A. Zhuravkov, Doctor of Physical and Mathematical Sciences, Y.V. Vasilevich, Doctor of Physical and Mathematical Sciences, A.V. Bogdanovich, Doctor of Engineering Sciences for their helpful comments that were allowed for in the present book.

References

1. Sherbakov, S.S., Sosnovskiy, L.A.: Mechanics of Tribo-Fatigue Systems. BSU Press, Minsk (2010). (in Russian)
2. Sosnovskiy, L.A., Sherbakov, S.S.: Surprises of Tribo-Fatigue. Magic Book, Minsk (2009)

3a. Sosnovskiy, L.A.: Tribo-Fatigue: Wear-Fatigue Damage and Its Prediction (Foundations of Engineering Mechanics). Springer, Berlin (2005)

3b. Sosnovskiy, L.A.: Fundamentals of Tribo-Fatigue. BelSUT Press, Gomel (2003). (in Russian)

3c. 索斯洛夫斯基著, L.A.: 摩擦疲劳学 磨损 – 疲劳损伤及其预测. 高万振译 – 中国矿业大学出版社 (2013) (in Chinese)

4. Sosnovskiy, L.A., Makhutov, N.A.: Methodological problems of a comprehensive assessment of damageability and limiting state of active systems. vol. 5, pp. 27–40, Factory Laboratory (1991)
5. Sosnovskiy, L.A.: Mechanics of Wear-Fatigue Damage. BelSUT Press, Go-mel (2007) (in Russian)

Chapter 2
Energy States of the Mechanothermodynamic System and the Analysis of Its Damageability

Abstract The fundamentals of the theory of evolution of mechanothermodynamic systems are developed using the energy concepts. The main feature of the theory is the analysis of damageability of material bodies due to absorption of (effective) energy caused by mechanical, thermodynamic loads, etc. The components of such energy are shown to interact dialectically. The theory of evolution of a system by damageability, as well as the fundamentals of the theory of limiting and translimiting states of a system is outlined. Particular cases of the theory of damageability are tested on experimental results.

2.1 General Notions

According to [18, 26, 37], the *mechanothermodynamic* (MTD) *system in the general case represents the thermodynamic continuum with solids distributed (scattered) within it, interacting with each other and with the continuum.* Consider its fragment of limited size $\Omega(X, Y, Z)$ shown in Fig. 2.1. The continuum has a temperature θ and a chemical composition Ch. Here there are two interacting solid elements (A and B) that can move relative to each other in the region of the contact area $S(x, y, z)$. Arbitrary mechanical loads applied to one of them (for example, to element A) in the x, y, z coordinate system can be reduced to the internal transverse forces Q_x, Q_y, Q_z, the longitudinal forces N_x, N_y, N_z and also to the bending moments M_x, M_y, M_z. Element B is pressed to element A by the loads that are reduced to the distributed normal pressure $p(x, y)$ and the tangential pressure $q(x, y)$. The origin of the coordinates is placed at the point of original contact O of the two elements (prior to volumetric deformation). It is easy to see that the elements A and B together form the *Tribo-Fatigue system* [18] which is the *friction pair* [37] *in the absence of internal forces* ($N_i = 0$, $Q_i = 0$, $M_i = 0$, $i = x, y, z$). Thus, the Tribo-Fatigue system is the *friction pair in which at least one of the elements supports non-contact loads* and, consequently, *undergoes volumetric deformation.* This representation of the MTD system has an advantage that the analysis of solid states and system components can adopt the appropriate solutions

L. Sosnovskiy and S. Sherbakov, *Mechanothermodynamics*,
DOI 10.1007/978-3-319-24981-0_2

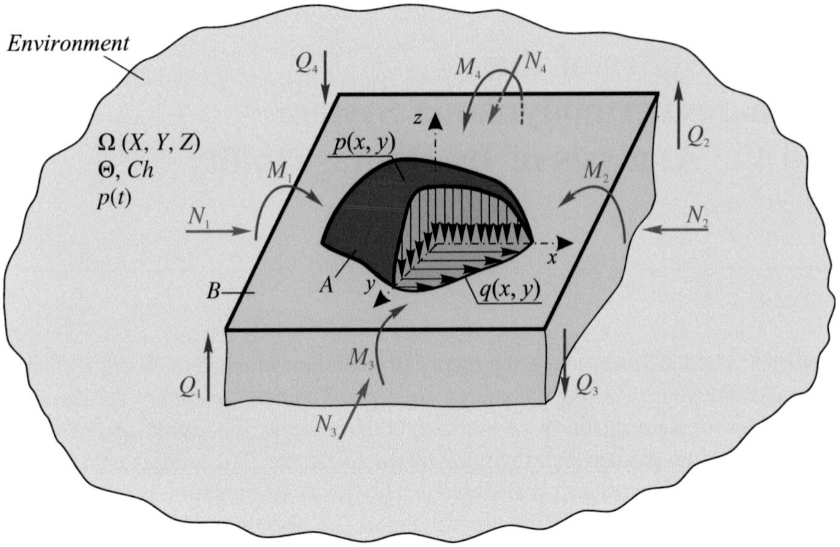

Fig. 2.1 Scheme of the elementary MTD system

known in deformable solid mechanics, in mechanics of contact interaction of solids, in mechanics of Tribo-Fatigue systems (Tribo-Fatigue), and in tribology.

Our main task is to describe the *energy state of the MTD system* under the action of mechanical and thermodynamic loads with regard to the environmental influence.

The energy state of any system is very interesting in itself. However, as applied to the MTD system it is very important to study *its damageability* and, as a result, to study the *conditions of reaching the limiting state.* Of special interest is the analysis of translimiting or supercritical conditions [37].

The main ideas, which are the fundamentals of the given theory, can be formulated considering [18, 30–32, 37] as follows.

I Due to the fact that the elements of the MTD system are subject to different-nature loads—mechanical, thermal, and electrochemical, the traditional analysis of their damageability and limiting state under the action only of mechanical stresses or strains [6, 7, 11, 13, 42, etc.] can be the basis for research. However this is not sufficient and, as a result, is ineffective. This means that there is a need to analyze MTD system states using more general —energy concepts.

II Considering that the damageability of solids of the MTD system is determined by mechanical, thermodynamic, and electrochemical loads, it is needed to introduce the *generalized representation of its complex damage* due to these loads acting at a time. Call such damage *any irreversible changes in shape, size, volume, mass, composition, structure, continuity* and, *as a result, physical-mechanical properties of the system elements.* This means the corresponding *changes in the functions of the system as a whole.*

III Generation and the development of complex damage are mainly determined by means of four *particular phenomena: mechanical fatigue, friction* and *wear, thermodynamic* and *electrochemical processes.* These phenomena are called particular phenomena in the sense that each of them can be realized as independent and separate. This leads to the corresponding energy state and damage in terms of particular (separate) criteria.

IV In the *general case,* all these particular phenomena and processes in the MTD system appear *simultaneously and within one zone.* The states of such a system are then caused not by one of any mentioned phenomena, but by *their joint (collective) development and, consequently, by their interaction.*

V Damages appear and develop not at one (dangerous) point of of a loaded solid with a working volume V_0, but within its dangerous volume $V_{ij} < V_0$ having a set of points, with each of which the *critical* level of stresses (strains) is achieved or surpassed with some probability.

VI If the *physical state* of the MTD system is described by its input energy U_Σ, then the *state of its damageability* is determined only by the *effective (dangerous) part* $U_\Sigma^{eff} \ll U_\Sigma$ that is spent for generation, motion, and interaction of irreversible damages.

VII The effective energy U_Σ^{eff} under volumetric and contanct deformation of solids can be represented by the function of three energy components: *thermal* U_T^{eff}, *force* U_n^{eff}, and *frictional* U_τ^{eff}:

$$U_\Sigma^{eff} = F_\Lambda\left(U_T^{eff}, U_{n(\sigma)}^{eff}, U_\tau^{eff}\right), \qquad (2.1)$$

where F_Λ takes into account the irreversible kinetic interaction of particular damage phenomena. The components $U_T^{eff}, U_n^{eff}, U_\tau^{eff}$ of the effective energy U_Σ^{eff} have no additivity.

VIII Processes of electrochemical (in particular, corrosion) damage of solids can be taken into consideration by introducing the parameter $0 \leq D_{ch} \leq 1$ and can be studied, for example, as *electrochemical damageability under the influence of temperature* ($D_{T(ch)}$), *stresses* ($D_{\sigma(ch)}$), *friction and corrosion* ($D_{\tau(ch)}$). So function (2.1) takes the form

$$U_\Sigma^{eff} = F_\Lambda\left(U_{T(ch)}^{eff}, U_{n(ch)}^{eff}, U_{\tau(ch)}^{eff}\right). \qquad (2.2)$$

IX The *generalized criterion of the limiting (critical) state* is represented by the condition when the effective energy u_Σ^{eff} reaches its limiting value—a *critical quantity* u_0 in some area of limited size—in the *dangerous volume* of the MTD system.

X Specific energy u_0 is considered to be a *fundamental constant for a given material*. It should not depend on testing conditions, input energy types, damage mechanisms.

XI In the general case, the limiting (critical) state of the MDT system is reached not due to a simple growth of effective energy components and, hence, due to the accumulation of irreversible damages caused by individual actions (loads) of different nature, but as a result of their *dialectical interaction, whose objectives are characterized by the development of phenomena of spontaneous hardening-softening of materials* in the given operating or testing conditions.

In such a way, taking into consideration function (2.2), the *hypothesis of the limiting (critical) state of the MTD system* can be represented in the following general form

$$\Phi(u_{n(ch)}^{eff}, u_{\tau(ch)}^{eff}, u_{T(ch)}^{eff}, \Lambda_{k\backslash l\backslash q}, m_k, u_0) = 0, \tag{2.3}$$

where the m_k's $k = 1, 2, \ldots$, are some characteristic properties (hardening-softening) of contacting materials, the $\Lambda_{k\backslash l\backslash n}$'s $\gtrless 1$ are the functions (parameters) of dialectical interactions of effective energies (irreversible damages) that are caused by different-nature loads. This means that at $\Lambda_k > 1$, the damageability increase is realized, at $\Lambda_l < 1$—its decrease, and at $\Lambda_n = 1$—its stable development.

XII Based on Item III, from the physical viewpoint, hypothesis (2.3) should be multicriterion, i.e., it should describe not only the states of the system as a whole, but its individual elements in terms of different criteria of performance loss (wear, fatigue damage, pitting, corrosion damage, thermal damage, etc.). In particular cases, it is possible to reach the corresponding limiting (critical) states in terms of one or two, three or several criteria at a time.

XIII The achievement of the limiting state

$$u_{\Sigma}^{eff} = u_0 \tag{2.4}$$

means a complete loss of the integrity of the MTD system, i.e., of all its functions. At the same time, the *damageability* of its elements

$$0 < \psi_u^{eff} = u_{\Sigma}^{eff} / u_0 \tag{2.5}$$

reaches the critical value

$$\psi_u^{eff}\left(\psi_{\sigma(ch)}, \psi_{\tau(ch)}, \psi_{T(ch)}, \Lambda_{k\backslash l\backslash q}, m_k\right) = 1. \tag{2.6}$$

XIV If $t = t_0$ is the time of origination of the system and $T_{\oplus}$ is the time when the system reaches its limiting state, then the *failure time of its functions* corresponds to the *relative life (longevity)* $t/T_{\oplus} = 1$. But the *lifetime* T_* *of the system* as the *material object* is longer than its existence time as a whole $(T_* \gg T_{\oplus})$ since at the time moment $t > T_{\oplus}$ the long process of system *degradation—disintegration* is realized by forming a great number of remains, pieces, fragments, etc. This process develops under the action of not only possible mechanical loads, but mainly under the environmental influence—up to the *system death as the material object* at the time moment $t = T_*$. The system death means its *complete disintegration into an infinitely large number of ultimately small particles* (for example, atoms). The *translimiting state of the system as a gradually disintegrating material object* can then be described by the following conditions

$$\psi_u^{eff} \to \infty; \tag{2.7}$$

$$d_\psi \to 0, \tag{2.8}$$

where d_ψ is the average size of disintegrating particles. At that, the organic relationship $\psi_u^{eff}(d_\psi)$ should exist between ψ_Σ and d_ψ. Then the *death condition* of the system is

$$t/T_* = 1. \tag{2.9}$$

XV Particles of "old system" disintegration are not destroyed, but are spent for the formation and growth of a number of "new systems". This is the essence of the *MTD system evolution hysteresis*.

2.2 Energy Theory of Limiting States

First, specify function (2.1).

To determine *speficic* effective energy, consider the *work of internal forces in an elementary volume dV of the Tribo-Fatigue system* ($A \leftrightarrows B$ in Fig. 2.1). In the general case, the differential of the work of the internal forces and the temperature dT_Σ can be written considering the rule of expanding the biscalar product of the *stress and strain tensors* σ and ε:

$$du = \sigma_{ij} \cdot d\varepsilon_{ij} + kdT_\Sigma = \begin{pmatrix} \sigma_{xx} & \sigma_{xy} & \sigma_{xz} \\ \sigma_{yx} & \sigma_{yy} & \sigma_{yz} \\ \sigma_{zx} & \sigma_{zy} & \sigma_{zz} \end{pmatrix} \cdot \begin{pmatrix} d\varepsilon_{xx} & d\varepsilon_{xy} & d\varepsilon_{xz} \\ d\varepsilon_{yx} & d\varepsilon_{yy} & d\varepsilon_{yz} \\ d\varepsilon_{zx} & d\gamma_{zy} & d\varepsilon_{zz} \end{pmatrix} + kdT_\Sigma$$
$$= \sigma_{xx}d\varepsilon_{xx} + \sigma_{yy}d\varepsilon_{yy} + \sigma_{zz}d\varepsilon_{zz} + \sigma_{xy}d\varepsilon_{xy} + \sigma_{xz}d\varepsilon_{xz} + \sigma_{yz}d\varepsilon_{yz} + kdT_\Sigma; \tag{2.10}$$

here k is the *Boltzmann* constant.

We proceed from the idea that in the general case, according to [13], the main role in forming wear-fatigue damage is played by the normal and shear stresses that cause the processes of *shear (due to friction)* and *tear (due to tension-compression).*

In this case, it is reasonable to divide the tensor σ into two parts: σ_τ is the *tensor of friction-shear stresses* or, briefly, the *shear tensor* and σ_σ is the *tensor of normal stresses (tension-compression),* or, briefly, the *tear tensor.* So in (2.10), the tear part σ_n and the shear part σ_τ of the tensor σ will be set as:

$$\begin{aligned} du &= \sigma_{ij}^{(V,W)} \cdot d\varepsilon_{ij}^{(V,W)} + kdT_\Sigma = \left(\sigma_n^{(V,W)} + \sigma_\tau^{(V,W)}\right) \cdot d\varepsilon_{ij}^{(V,W)} + kdT_\Sigma \\ &= \sigma_n^{(V,W)} \cdot d\varepsilon_{ij}^{(V,W)} + \sigma_\tau^{(V,W)} \cdot d\varepsilon_{ij}^{(V,W)} + kdT_\Sigma = du_n + du_\tau + du_T. \end{aligned} \tag{2.11}$$

According to Items III and IV, the tensors σ and ε should be represented as follows:

$$\sigma_{ij} = \sigma_{ij}^{(V,W)} = \sigma_{ij}\left(\sigma_{ij}^{(V)}, \sigma_{ij}^{(W)}\right), \quad \varepsilon_{ij} = \varepsilon_{ij}^{(V,W)} = \varepsilon_{ij}\left(\varepsilon_{ij}^{(V)}, \varepsilon_{ij}^{(W)}\right). \tag{2.12}$$

Here, the stress and strain tensors with the superscript V are caused by the action of volumetric loads (the general cases of 3D bending, torsion, and tension-compression) and those with the superscript W—by the contact interaction of the system elements.

Expression (2.11) with regard to (2.12) can be given as follows:

$$\begin{aligned} du &= \sigma_{ij}^{(V,W)} \cdot d\varepsilon_{ij}^{(V,W)} + kdT_\Sigma = \left(\sigma_n^{(V,W)} + \sigma_\tau^{(V,W)}\right) \cdot d\varepsilon_{ij}^{(V,W)} + kdT_\Sigma \\ &= \sigma_n^{(V,W)} \cdot d\varepsilon_{ij}^{(V,W)} + \sigma_\tau^{(V,W)} \cdot d\varepsilon_{ij}^{(V,W)} + kdT_\Sigma = du_n + du_\tau + du_T. \end{aligned} \tag{2.13}$$

In the case of the linear relationship between the stresses and strains, expression (2.12) will assume the form:

$$\sigma_{ij} = \sigma_{ij}^{(V,W)} = \sigma_{ij}^{(V)} + \sigma_{ij}^{(W)} = \begin{pmatrix} \sigma_{xx}^{(V)} + \sigma_{xx}^{(W)} & \sigma_{xy}^{(V)} + \sigma_{xy}^{(W)} & \sigma_{xz}^{(V)} + \sigma_{xz}^{(W)} \\ \sigma_{yx}^{(V)} + \sigma_{yx}^{(W)} & \sigma_{yy}^{(V)} + \sigma_{yy}^{(W)} & \sigma_{yz}^{(V)} + \sigma_{yz}^{(W)} \\ \sigma_{zx}^{(V)} + \sigma_{zx}^{(W)} & \sigma_{zy}^{(V)} + \sigma_{zy}^{(W)} & \sigma_{zz}^{(V)} + \sigma_{zz}^{(W)} \end{pmatrix}, \tag{2.14}$$

$$\varepsilon_{ij} = \varepsilon_{ij}^{(V,W)} = \varepsilon_{ij}^{(V)} + \varepsilon_{ij}^{(W)} = \begin{pmatrix} \varepsilon_{xx}^{(V)} + \varepsilon_{xx}^{(W)} & \varepsilon_{xy}^{(V)} + \varepsilon_{xy}^{(W)} & \varepsilon_{xz}^{(V)} + \varepsilon_{xz}^{(W)} \\ \varepsilon_{yx}^{(V)} + \varepsilon_{yx}^{(W)} & \varepsilon_{yy}^{(V)} + \varepsilon_{yy}^{(W)} & \varepsilon_{yz}^{(V)} + \varepsilon_{yz}^{(W)} \\ \varepsilon_{zx}^{(V)} + \varepsilon_{zx}^{(W)} & \varepsilon_{zy}^{(V)} + \varepsilon_{zy}^{(W)} & \varepsilon_{zz}^{(V)} + \varepsilon_{zz}^{(W)} \end{pmatrix}, \tag{2.15}$$

and (2.13) will be as follows:

$$du = u = \frac{1}{2}\sigma_{ij}\cdot\varepsilon_{ij} + kT_\Sigma = \frac{1}{2}\left(\sigma_{ij}^{(V)} + \sigma_{ij}^{(W)}\right)\cdot\left(\varepsilon_{ij}^{(V)} + \varepsilon_{ij}^{(W)}\right) + kT_\Sigma$$
$$= \frac{1}{2}\left[\left(\sigma_n^{(V)} + \sigma_n^{(W)}\right) + \left(\sigma_\tau^{(V)} + \sigma_\tau^{(W)}\right)\right]\cdot\left(\varepsilon_{ij}^{(V)} + \varepsilon_{ij}^{(W)}\right) + kT_\Sigma$$
$$= \frac{1}{2}\left[\begin{pmatrix} \sigma_{xx}^{(V)} + \sigma_{xx}^{(W)} & 0 & 0 \\ 0 & \sigma_{yy}^{(V)} + \sigma_{yy}^{(W)} & 0 \\ 0 & 0 & \sigma_{zz}^{(V)} + \sigma_{zz}^{(W)} \end{pmatrix} + \right.$$
$$\left. + \begin{pmatrix} 0 & \sigma_{xy}^{(V)} + \sigma_{xy}^{(W)} & \sigma_{xz}^{(V)} + \sigma_{xz}^{(W)} \\ \sigma_{yx}^{(V)} + \sigma_{yx}^{(W)} & 0 & \sigma_{yz}^{(V)} + \sigma_{yz}^{(W)} \\ \sigma_{zx}^{(V)} + \sigma_{zx}^{(W)} & \sigma_{zy}^{(V)} + \sigma_{zy}^{(W)} & 0 \end{pmatrix}\right] \tag{2.16}$$
$$\cdot \begin{pmatrix} \varepsilon_{xx}^{(V)} + \varepsilon_{xx}^{(W)} & \varepsilon_{xy}^{(V)} + \varepsilon_{xy}^{(W)} & \varepsilon_{xz}^{(V)} + \varepsilon_{xz}^{(W)} \\ \varepsilon_{yx}^{(V)} + \varepsilon_{yx}^{(W)} & \varepsilon_{yy}^{(V)} + \varepsilon_{yy}^{(W)} & \varepsilon_{yz}^{(V)} + \varepsilon_{yz}^{(W)} \\ \varepsilon_{zx}^{(V)} + \varepsilon_{zx}^{(W)} & \varepsilon_{zy}^{(V)} + \varepsilon_{zy}^{(W)} & \varepsilon_{zz}^{(V)} + \varepsilon_{zz}^{(W)} \end{pmatrix} + kT_\Sigma.$$

From (2.16) it is seen that the tear part σ_n of the tensor σ is the sum of the tear parts of the tensors under the volumetric deformation $\sigma_n^{(V)}$ and the surface load (friction) $\sigma_n^{(W)}$, whereas the shear part σ_τ is the sum of the shear parts $\sigma_\tau^{(V)}$ and $\sigma_\tau^{(W)}$. This means the *fundamental particularity* of the generalized approach to the construction of the criterion for the limiting state of the MTD system.

The effective part of total energy (2.16) is separated according to Items V and VIII with regard to [30–32, 37]. To do this, introduce the coefficients of appropriate dimensions $A_n(V)$, $A_\tau(V)$, and $A_T(V)$ that determine the fraction of the absorbed energy

$$du_\Sigma^{eff} = \Lambda_{M\setminus T}(V)\left\{\Lambda_{n\setminus\tau}(V)\left[A_n(V)\sigma_n\cdot d\varepsilon_{ij} + A_\tau(V)\sigma_\tau\cdot d\varepsilon_{ij}\right] + A_T(V)kdT_\Sigma\right\}, \tag{2.17}$$

or

$$du_\Sigma^{eff} = \Lambda_{M\setminus T}(V)\left\{\Lambda_{n\setminus\tau}(V)\left[A_n(V)du_n + A_\tau(V)du_\tau\right] + A_T(V)du_T\right\}, \tag{2.18}$$

where $\Lambda_{M\setminus T}(V)$ and $\Lambda_{\tau\setminus\sigma}(V)$ are the interaction functions of different-nature energies. The subscript $\tau\setminus\sigma$ means that the function Λ describes the interaction between the shear (τ) and tear (σ) components of effective energy, and the subscript $M\setminus T$ means that the function Λ describes the interaction between the mechanical (M) and thermal (T) parts of effective energy. The fact that generally speaking, the coefficients A can be different for different points of the volume V, enables one to allow for the continuum inhomogeneity. Taking into consideration (2.18), criteria (2.3) can be specified with no regard to the influence of the electrochemical properties (ch) of the environment:

$$\Lambda_{M\backslash T}(V)\left\{\Lambda_{n\backslash\tau}(V)\left[u_n^{eff}+u_\tau^{eff}\right]+u_T^{eff}\right\}=u_0. \quad (2.19)$$

In the case of the linear relationship between the stresses and strains, expressions (2.17) and (2.18) will be as follows:

$$u_\Sigma^{eff}=\Lambda_{M\backslash T}(V)\left\{\Lambda_{n\backslash\tau}(V)\left[\frac{1}{2}A_n(V)\,\sigma_n\cdot\varepsilon_{ij}+\frac{1}{2}A_\tau(V)\,\sigma_\tau\cdot\varepsilon_{ij}\right]+A_T(V)\,kT_\Sigma\right\}, \quad (2.20)$$

or

$$\begin{aligned}u_\Sigma^{eff}&=\Lambda_{M\backslash T}(V)\{\Lambda_{n\backslash\tau}(V)\,[A_n(V)u_n(V)+A_\tau(V)u_\tau(V)]+A_T(V)u(V)\}\\&=\Lambda_{M\backslash T}(V)\left\{\Lambda_{n\backslash\tau}(V)\,\left[u_n^{eff}(V)+u_\tau^{eff}(V)\right]+u_T^{eff}(V)\right\}.\end{aligned} \quad (2.21)$$

With regard to expression (2.12), criterion (2.19) can be represented as follows:

$$u_\Sigma^{eff}=\left\{\left[u_n^{eff}(\sigma_n^{(V,W)},\varepsilon_n^{(V,W)})+u_\tau^{eff}(\sigma_\tau^{(V,W)},\varepsilon_\tau^{(V,W)})\right]\Lambda_{n\backslash\tau}+u_T^{eff}\right\}\Lambda_{T\backslash M}=u_0. \quad (2.22)$$

When the time effects should be taken into consideration, criterion (2.22) will assume the form:

$$u_{\Sigma t}^{eff}=\int_0^t\left\{\left[u_n^{eff}(\sigma_n^{(V,W)},\varepsilon_n^{(V,W)},t)+u_\tau^{eff}(\sigma_\tau^{(V,W)},\varepsilon_\tau^{(V,W)},t)\right]\Lambda_{n\backslash\tau}(t)+u_T^{eff}(t)\right\}\Lambda_{T\backslash M}(t)\,dt=u_0. \quad (2.23)$$

Thus, expression (2.21) is a concrete definition of function (2.1) and formula (2.22) is a concrete definition of criterion (2.3) for that case when the electrochemical influence of the environment is not allowed for.

Criterion (2.3) in forms (2.22) and (2.23) says: *when the sum of interacting effective energy components when acted upon by force, frictional, and thermal (thermodynamic) loads reach the critical (limiting) quantity* u_0, *the limiting (or critical) state of the MTD system (of the both individual elements of the system and the system as a whole) is realized. Physically, this state is determined by many and different damages.*

The fundamental character of the parameter u_0 has been mentioned above. According to [16, 48, 49], the parameter u_0 will be interpreted as the *initial activation energy of the disintegration process.* It is shown that the quantity u_0 approximately *corresponds both to the sublimation heat for metals* and *crystals with ionic bonds* and *to the activation energy of thermal destruction for polymers,* i.e.,:

$$u_0 \approx u_T.$$

On the other hand, the quantity u_0 is determined as the *activation energy for mechanical failure:*

$$u_0 \approx u_M.$$

In such a way, the energy u_0 can be considered to be the *material constant*:

$$u_0 \approx u_M \approx u_T = \text{const}. \tag{2.24}$$

Taking into consideration the physical-mechanical and thermodynamic representations of the processes of damageability and failure [5, 8, 48], write down (2.24) in the following form:

$$u_M = s_k \frac{\sigma_{th}}{E} \frac{C_a}{\alpha_V} = u_0 = kT_S \ln \frac{k\theta_D}{h} = u_T, \tag{2.25}$$

where s_k is the reduction coefficient, σ_{th} is the theoretical strength, E is the elasticity modulus, C_a is the heat capacity of atom, α_V is the thermal expansion of volume, k is the *Boltzmann* constant, T_S is the melting point, θ_D is the *Debye* temperature, h is the *Planck* constant. According to (2.25), it can be taken approximately [48]

$$u_0 \approx \varepsilon_* \frac{C_a}{\alpha_V}, \tag{2.26}$$

where $\varepsilon_* \approx 0.6$ is the ultimate strain of the interatomic bond. Calculations according to (2.26) are not difficult. Methods of experimental determination of the quantity u_0 have also been developed [16].

From equality (2.25), it follows that u_0 is the activation energy of a given material, which is by the order of magnitude equal to 1–10 eV per one particle or molecule ($\sim 10^2$ to 10^3 kJ/mol), i.e., *the value that is close to the energy of interatomic bond rupture in the solid* [2]. Its level does not depend on how the rupture is reached—mechanically, thermally or by their simultaneous action. In [16], it is possible to find the tables containing the u_0 values for different materials.

From (2.25), it is possible to find the *thermomechanical constant of material* [37]

$$\frac{\sigma_{th}}{T_S} = E \frac{\alpha_V k}{C_a} \ln \frac{k\theta_D}{h} = \theta_\sigma. \tag{2.27}$$

The constant θ_σ characterizes the *strength loss per 1 K*.

2.3 Energy Theory of Damage

Criterion (2.22) is written in absolute values of physical parameters—in values of effective and critical energy components. This criterion can be easily made dimensionless by diving it by the quantity u_0. It can then be represented *in terms of irreversible (effective) damage*

$$\psi_u^{eff} = \frac{u_\Sigma^{eff}}{u_0} = 1. \tag{2.28}$$

It is clear that the *local (at the point) energy measure of damageability* ψ_u^{eff} varies within the range

$$0 \leq \psi_u^{eff} \leq 1, \tag{2.29}$$

or in detailed form

$$0 \leq \psi_u^{eff} = \frac{\Lambda_{T\backslash M}}{u_0}\left\{ \left[u_n^{eff}(\sigma_n^{(V,W)}, \varepsilon_n^{(V,W)}) + u_\tau^{eff}(\sigma_\tau^{(V,W)}, \varepsilon_\tau^{(V,W)})\right]\Lambda_{n\backslash\tau} + u_T^{eff}\right\} \leq 1. \tag{2.29a}$$

According to (2.29a), *particular energy measures of damageability* can also be determined:

$$0 \leq \psi_n^{eff} = \frac{u_n^{eff}\left(\sigma_n^{(V,W)}, \varepsilon_n^{(V,W)}\right)}{u_0} \leq 1; \tag{2.30}$$

$$0 \leq \psi_\tau^{eff} = \frac{u_\tau^{eff}\left(\sigma_\tau^{(V,W)}, \varepsilon_\tau^{(V,W)}\right)}{u_0} \leq 1; \tag{2.31}$$

$$0 \leq \psi_T^{eff} = \frac{u_T^{eff}}{u_0} \leq 1 \tag{2.32}$$

due to effective different-nature energies that are determined by force (subscripts n), frictional (subscripts τ), and thermodynamic (subscripts T) loads, respectively. Now criterion (2.28) can be written in dimensionless form

$$\psi_u^{eff} = \left[(\psi_n^{eff} + \psi_\tau^{eff})\Lambda_{n\backslash\tau} + \psi_T^{eff}\right]\Lambda_{M\backslash T} = 1. \tag{2.33}$$

According to (2.33), the limiting state of the MTD system is reached when the sum of interacting damages $(0 < \psi < 1)$ due to mechanical and thermodynamic loads becomes equal to 1. Criterion (2.22) in form (2.33) is convenient because all damageability measures are dimensionless and are within the same range $0 < \psi < 1$.

Since numerous and infinite actions and the interaction effects of physical damages of many types (dislocations, vacancies, inelastic deformations, etc.) cannot be described and predicted exactly, when analyzing the MTD system, one introduces the concept of the *interaction of dangerous volumes* [37] that contain a real complex of damages [defects generated by the action of the corresponding fields of stresses (strains)]. According to the *statistical model of the deformable solid with the dangerous volume* [38], for example, in the case of fatigue damage of the construction element in the linear stress state, the volume should depend on its geometric parameters responsible for the working volume V_0 of the solid, on distribution function parameters $p(\sigma_{-1})$ and $p(\sigma)$ of the fatigue limit σ_{-1} and on acting stresses σ considering both the probabilities P and γ_0, as well as gradients G_σ of acting stresses:

$$V_{P_\gamma} = F_V[p(\sigma_{-1}), p(\sigma), G_\sigma, V_0, P, \gamma_0, \vartheta_V]. \tag{2.34}$$

Here, ϑ_V describes how the fatigue limit is influenced by the shape of the solid and the scheme of its loading in fatigue tests.

Thus, the *dangerous volume can serve as the equivalent of the complex of damages,* as its value is proportional, in particular to the level of effective stresses and, hence, to the number (concentrations) of defects (damages).

As follows from expression (2.34), the boundary between dangerous and safe volumes is generally blurred and probabilistic in nature. As the damage probability P of the solid increases, the dangerous volume V_{P_γ} is growing. At a given value of P, the volume can vary depending on the confidence probability γ_0. It means that at P = const

$$V_{P_{\gamma\,\min}} \le V_{P_\gamma} \le V_{P_{\gamma\,\max}}, \tag{2.35}$$

if $\gamma_{\min} \le \gamma_0 \le \gamma_{\max}$. Here, $\gamma_{\min}, \gamma_{\max}$ form the permissible range of γ_0. If it is assumed that $\gamma_0 = \text{const}$, then the dangerous volume will have a single value associated with the damage probability P.

Scattered damage within the dangerous volume is characteristic not only for the so-called smooth solids but also for the elements with structural *stress concentrators* [38]. Figure 2.2 demonstrates several microcracks in the sharp notch (rounding-off radius r = 0.5 mm, the theoretical stress concentration factor $\alpha_n = 8$ in Fig. 2.2a) and in the flat notch (r = 2 mm, $\alpha_n = 2.55$ in Fig. 2.2b) and also two fatigue cracks at a distance of 25 mm from each other at the filet joint from the crankshaft journal to its web (r = 18 mm, $\alpha_n = 3.2$ in Fig. 2.2c); the crankshaft journal diameter is 360 mm.

Thus, if in the uniaxial stress state, the stress distribution σ (x, y, z) in the x, y, z coordinates is known, then the *dangerous volume should be calculated by the formula*

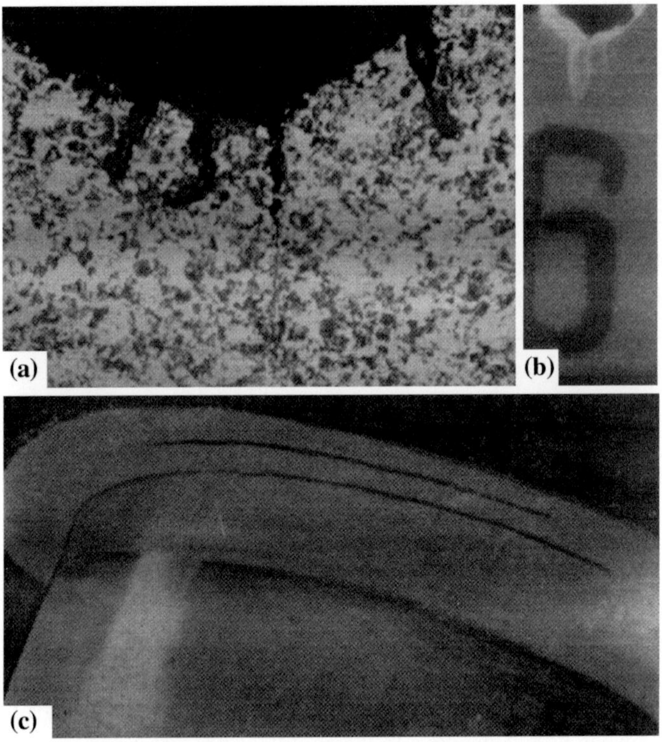

Fig. 2.2 Fatigue microcracks in the zones of stress concentrators (L.A. Sosnovskiy)

$$V_{P_\gamma} = \iiint\limits_{\sigma(x,y,z)\,>\,\sigma_{-1\min}} dxdydz, \tag{2.36}$$

Here $\sigma_{-1\min}$ being the lower boundary of the range of the fatigue limit σ_{-1} is such that if $\sigma_{-1} < \sigma_{-1\,\min}$, then $P = 0$.

From expression (2.36) it follows that the *generalized condition for fatigue failure* is of the form

$$V_{P_\gamma} > 0 \tag{2.37}$$

with some probability P under the confidence probability γ_0.

If

$$V_{P_\gamma} = 0, \tag{2.38}$$

then physically, fatigue failure cannot occur (as in this case, $\sigma < \sigma_{-1\min}$); hence, (2.38) is the *generalized condition of no-failure.*

The methods for calculation of dangerous volumes V_{ij} for friction pairs and Tribo-Fatigue systems are developed similar to (2.34)

$$V_{ij} = V_{ij}\left(\sigma_n^{(V,W)}, \sigma_\tau^{(V,W)}, \sigma_{\lim}^{(V,W)}, G_{\sigma_{ij}}, V_0, P, \gamma_0\right) \tag{2.39}$$

and outlined elsewhere in [18, 19, 46, 47]. Here, $\sigma_{\lim}^{(V,W)}$ is the limiting stress based on an assigned criterion of damage and failure. Further, the following dimensionless damageability characteristics can be introduced: *integral energy damageability within the dangerous volume*

$$\Psi_u^{eff}(V) = \iiint\limits_{\psi_u^{eff}(dV)\geq 1} \frac{u_\Sigma^{eff}}{u_0} dV \tag{2.40}$$

and *the mean energy damageability (at each point of the dangerous volume)*

$$\bar{\Psi}_u^{eff}(V) = \frac{1}{V_u} \iiint\limits_{\psi_u^{eff}(dV)\geq 1} \frac{u_\Sigma^{eff}}{u_0} dV. \tag{2.41}$$

The *accumulation of energy damageability in time within the dangerous volume* is described by the formulas

$$\Psi_u^{eff}(V,t) = \int\limits_t \iiint\limits_{\psi_u^{eff}(dV)\geq 1} \frac{u_\Sigma^{eff}}{u_0} dV dt, \tag{2.42}$$

$$\bar{\Psi}_u^{eff}(V,t) = \int\limits_t \frac{1}{V_u} \iiint\limits_{\psi_u^{eff}(dV)\geq 1} \frac{u_\Sigma^{eff}}{u_0} dV dt. \tag{2.43}$$

Having (2.38)–(2.43), the damageability of the MTD system can be described and analyzed using the most general representations—energy concepts with regard to the influence of numerous and different factors taken into account by (2.34), including the *scale effect,* i.e., the changes in the size and shape (mass) of system elements.

According to [1, 37], the function $\Lambda_{k/l/n}$ for damage interactions in the MTD system is determined by the parameters ρ of the effective energy ratio:

$$\Lambda_{k\backslash l\backslash q} = \Lambda_{k\backslash l\backslash q}\left(\rho_{M\backslash T}, \rho_{n\backslash\tau}\right) \gtreqless 1; \tag{2.44}$$

$$\rho_{n\backslash\tau} = u_\tau^{eff}/u_n^{eff}, \quad \rho_{M\backslash T} = u_M^{eff}\Big/u_T^{eff}. \tag{2.45}$$

The quantities Λ calculated by (2.44) describe the influence of the level of load parameter ratios on the character and direction of interaction of irreversible damages. *If* $\Lambda > 1$, then the *system is self-softening* because at balance of hardening-softening phenomena, softening processes are dominant. *If* $\Lambda < 1$, then the *system is self-hardening,* because at balance of hardening-softening phenomena, hardening processes are dominant. At $\Lambda = 1$, the *system appears to be stable*—the *spontaneous hardening-softening phenomena are at balance within it.* The general analysis of damage interactions in the MTD systems will be given below (see Chap. 3).

2.4 Account of Electrochemical Damageability

After criterion (2.3) has been formalized in principle, the action of electrochemical loads (damages) should be taken into consideration according to Item VII. It should be said at once that this is difficult to perform in the strict mechanothermodynamic statement: electrochemical reactions at environment-deformable solid interactions are very diverse, complex and are insufficiently studied. That's why, the approach proposed in [30–32, 37] is adopted here: the simplification is introduced, according to which the damageability of solids in the environment is determined by corrosion-electrochemical processes. In addition, the *hypothesis is put forward, following which the effective energy of corrosion-electrochemical damage is proportional to the square of the corrosion speed*, i.e.,

$$u_{ch}^{eff} \sim v_{ch}^2. \tag{2.46}$$

If according to Item VII, $0 \le D_{ch} \le 1$ is the parameter of corrosion-electrochemical damage of the solid, then based on [28], criterion (2.2) with regard to (2.22) will be as follows:

$$\Lambda_{M\backslash T}\left[\left(\frac{u_n^{eff}\left(\sigma_n^{(V,W)}, \varepsilon_n^{(V,W)}\right)}{u_0(1-D_n)} + \frac{u_\tau^{eff}\left(\sigma_\tau^{(V,W)}, \varepsilon_\tau^{(V,W)}\right)}{u_0(1-D_\tau)}\right)\Lambda_{n\backslash\tau} + \frac{u_T^{eff}}{u_0(1-D_T)}\right] = 1,$$
$$\Lambda \gtreqless 1, \tag{2.47}$$

where

$$0 \le \frac{u_n^{eff}\left(\sigma_n^{(V,W)}, \varepsilon_n^{(V,W)}\right)}{u_0(1-D_n)} = \psi_{n(ch)}^{eff} \le 1; \tag{2.48}$$

$$0 \leq \frac{u_{\tau}^{eff}\left(\sigma_{\tau}^{(V,W)}, \varepsilon_{\tau}^{(V,W)}\right)}{u_0(1 - D_{\tau})} = \psi_{\tau(ch)}^{eff} \leq 1; \tag{2.49}$$

$$0 \leq \frac{u_T^{eff}}{u_0(1 - D_T)} = \psi_{T(ch)}^{eff} \leq 1; \tag{2.50}$$

$$1 - D_T = b_{e(T)}\left(\frac{v_{ch}}{v_{ch(T)}}\right)^{m_{v(T)}}; 1 - D_n = b_{e(n)}\left(\frac{v_{ch}}{v_{ch(n)}}\right)^{m_{v(n)}}; \quad 1 - D_{\tau} = b_{e(\tau)}\left(\frac{v_{ch}}{v_{ch(\tau)}}\right)^{m_{v(\tau)}}, \tag{2.51}$$

where v_{ch} is the corrosion speed in this environment, $v_{ch(T)}$, $v_{ch(\sigma)}$, $v_{ch(\tau)}$ is the corrosion speed in the same environment under thermal, force, and friction actions, respectively; the b_e's are the coefficients responsible for corrosive erosion processes; the $m_{V(\cdot)}$'s are the parameters responsible for the electrochemical activity of materials at force (the subscript σ), friction (the subscript τ), and thermodynamic (the subscript T) loads, wherein $m_{V(\cdot)} = 2/A_{ch}$ and the parameter $A_{ch} \gtreqless 1$.

In [12], other methods for assessment of the parameter D_{ch} can be found.

As seen, Eq. (2.47) is a specific definition of criterion (2.3). According to this criterion, the *limiting state of the MTD system is reached when the sum of dialectically interacting effective damages due to force, friction, and thermodynamic loads (including electrochemical damage when acted upon by stress, friction, temperature) becomes equal to unity.*

2.5 Some Special Cases

Further, consider the specific case when in (2.21) it is assumed that $A_\sigma(V) = A_\sigma$ = const, $A_\tau(V) = A_\tau$ = const, $A_T(V) = A_T$ = const, $A_{\tau\backslash\sigma}(V) = A_{\tau\backslash\sigma}$ = const, $A_{M\backslash T}(V) = A_{M\backslash T}$ = const.

In this case, firstly, the stress state is caused by volumetric deformation, for which all stress tensor components, except one component σ (one-dimensional tension-compression, pure bending), can be neglected. Secondly, the stress state is caused by surface friction, for which all stress tensor components, except one component τ_w, can be ignored. Then (2.21) assumes the following form:

$$\Lambda_{M\backslash T}\left[\Lambda_{n\backslash\tau}\left(A_n\sigma^2 + A_\tau\tau^2\right) + A_T T_\Sigma\right] = u_0,$$

or in accordance with (2.47)

$$\Lambda_{M\backslash T}\left[\frac{a_T}{1 - D_T}T_\Sigma + \Lambda_{n\backslash\tau}\left(\frac{a_n}{1 - D_n}\sigma^2 + \frac{a_\tau}{1 - D_\tau}\tau_w^2\right)\right] = u_0, \quad \Lambda \gtreqless 1, \tag{2.52}$$

where $\frac{a_n}{1-D_n} = A_n$, $\frac{a_\tau}{1-D_\tau} = A_\tau$, $\frac{a_T}{1-D_T} = A_T$.

Thus, Eq. (2.52) is the simplest form of the energy criterion of the limiting state that is nevertheless of great practical importance [37].

If there is no electrochemical influence of the environment ($D_{ch} = 0$), then

$$u_{\Sigma}^{eff} = \Lambda_{M\backslash T}\left[a_T T_{\Sigma} + \Lambda_{n\backslash\tau}\left(a_n\sigma^2 + a_{\tau}\tau_w^2\right)\right] = u_0, \quad \Lambda \gtreqless 1 \tag{2.53}$$

Equation (2.53) is the simplest form of the energy criterion of the limiting state that is of great practical importance [1, 37, 43]. It serves particularly for the development of methods of assessing the parameters $a_T, a_{\sigma}, a_{\tau}$. In fact, at $\Lambda_{M\backslash T} = \Lambda_{\tau\backslash n} = 1$, the boundary conditions are the following:

$$\left.\begin{array}{lll} T_{\Sigma} = 0,\ \tau_w = 0: & a_n\sigma_d^2 = u_0, & a_n = u_0/\sigma_d^2; \\ T_{\Sigma} = 0,\ \sigma = 0: & a_{\tau}\tau_d^2 = u_0, & a_{\tau} = u_0/\tau_d^2; \\ \sigma = 0,\ \tau_w = 0: & a_T\sigma_d = u_0, & a_T = u_0/T_d, \end{array}\right\} \tag{2.54}$$

where σ_d, τ_d are the normal and friction limiting stresses as $T \to 0$. These are called (mechanical) destruction limits, T_d is the destruction temperature (when $\sigma = 0$, $\tau_w = 0$) or the thermal destruction limit [28].

The effective ("dangerous") part of total strain energy can also be determined from the following physical considerations. It shall be assumed that the *strain energy flow u* generated in the material sample subject to its strain cycling ($\varepsilon = \varepsilon_{\max}\sin\omega t$) in the homogeneous (linear) stress state is to a certain extent *similar to the light flux*. In fact, it is continuously excited when the loading cycle is repeated at the frequency $\omega = 1/\lambda$. This permits one to consider it as a wave (with a length λ). Some part of the energy u generated in such a way can be absorbed by material atoms and structural formations, followed by the material damage. Denote the absorbed part of the energy as u^{eff}. The generated energy u is then equal to:

$$u = u^{eff} + u_{cons}, \tag{2.55}$$

Here, u_{cons} is the non-absorbed part (here it is called the conservative part) of the generated energy u.

If the analogy of light and strain energy is justified, then the strain absorption law may be similar to Bouguer's light absorption law. Consequently, the equation relating the energy u_{cons} passed through the deformed material volume V and the generated energy u is of the form:

$$u_{cons} = u\exp(-\chi_{\varepsilon}V), \tag{2.56}$$

or, in accordance with *Lambert*, in differential form:

$$\frac{du}{u} = -\chi_{\varepsilon}V. \tag{2.57}$$

Here, as in *Bourguer-Lambert's* equation, the coefficient χ_ε independent of u is the energy absorption parameter.

Taking into account (2.56) in (2.55), the *absorption law* of *strain energy* is obtained:

$$u^{eff} = u[1 - \exp(-\chi_\varepsilon V)], \tag{2.58}$$

and, hence, if $u = 0$ or $V = 0$, then $u^{\text{eff}} = 0$. If $V \to \infty$, then it appears that according to (2.56), $u_{cons} = u$, i.e., all supplied energy is dissipated within such a volume.

From the physical point of view, the strain energy absorption process is caused by many phenomena:

- electron transition in absorbing atoms from lower to higher energy levels (quantum theory [9]);
- generation and development of dislocation structures (dislocation theory [4]);
- emergence of II and III order residual strains (stresses) (elasticity theory [41]);
- formation and development of any imperfections (defects) of material composition and structure—point, planar, and spatial (physical materials science [20]);
- hardening-softening phenomena (including strain aging) developing in time (fatigue theory [43]);
- changes in (internal) Tribo-Fatigue entropy (wear-fatigue damage mechanics [37]).

It should be noted that approach (2.58) can also be extended to the case of friction, since any indenter drives a strain wave upstream in the thin surface layer of the solid to which the indenter is pressed to [40]. In this case, χ_γ will be the energy absorption parameter. The subscript γ denotes the shear strain. Similarly, heat absorption in the deformable solid can also be considered. Finally, the problem of strain energy absorption in the inhomogeneous (including complex) stress state can be easily solved by setting the dangerous volume $V = V_{P_\gamma}$ into (2.56)–(2.58).

It should be noted that *although criterion* (2.53) *is particular, it is fundamental and general in nature.* Its general nature is caused by the fact that in this case, all four particular phenomena responsible for the MTD system state (in the statement simplified in terms of the stress-strain state) are taken into account (in accordance with Item III). Its fundamental nature is that here, as in complete solution (2.21), $\Lambda_{n\backslash\tau}$ takes into account the interaction of mechanical components of effective energy due to friction τ_w and normal σ stresses, whereas $\Lambda_{M\backslash T}$ takes into account the interaction of thermal and mechanical components of effective energy. The thermal component of effective energy is determined through the variations of the total temperature $T_\Sigma = T_2 - T_1$ in the zone of force contact caused by all sources of heat, including the one released during mechanical (spatial and surface) deformation, structural changes, etc.

From (2.53) it is easy to obtain a number of formulas important for application. So, the conditions of purely thermal (or thermodynamic) failure (when $\sigma = 0$ and $\tau_w = 0$) or purely mechanical failure (when $T_\Sigma \to 0$) will be as follows:

$$a_T T_\Sigma = u_0; \tag{2.59}$$

$$\Lambda_{n\backslash\tau}\left(a_n\sigma^2 + a_\tau\tau_w^2\right) = u_0. \tag{2.60}$$

For *isothermal mechanical fatigue* (when $\tau_w = 0$), we have

$$\Lambda_{M\backslash T}\left(a_T T_\Sigma + a_n\sigma^2\right) = u_0, \tag{2.61}$$

and for *isothermal frictional fatigue* (when $\sigma = 0$), we obtain

$$\Lambda_{M\backslash T}\left(a_T T_\Sigma + a_\tau\tau_w^2\right) = u_0. \tag{2.62}$$

The general analysis of the above-described partial criteria allows three main conclusions to be made.

(1) The growth of load parameters (σ, τ_w, T_Σ, D) results in a corresponding acceleration to achieve the limiting state (u_0).
(2) The limiting state of the system can also be reached by increasing only one (any) of the load parameters (other parameters remain unchanged).
(3) If $\Lambda > 1$, then the damageability of the system accordingly enhances (i.e., the processes of its softening are dominant), and if $\Lambda < 1$, then it slows down (i.e., the processes of its hardening are advantageous) in comparison with the damageability due to the joint action of load parameters alone (with no regard to the dialectical interaction of irreversible damages).

The last conclusion is also the result of a fundamentally new approach to the construction of the criterion of the limiting state of MTD systems [22]. According to this approach, *nonreciptocal influence of factors, but the interaction* ($\Lambda \gtreqless 1$) *of phenomena determines the damageability processes in the MTD system* [22, 25–27, 45]. In this regard, the results of more than 600 diverse experimental data were analyzed and synthesized. This permitted the generalized MTD function of states critical for damageability to be revealed.

2.6 Analysis of Experimental Results

Experimental analysis of generalized criterion (2.47) of the limiting state of a MTD system is extremely difficult as there are no relevant experimental data. Data acquisition is though very relevant, but at the same time is very difficult and expensive. Therefore, the analysis of particular criterion (2.61) is given below.

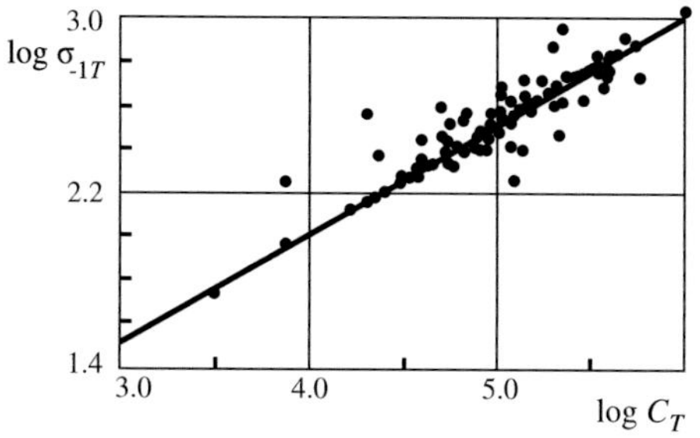

Fig. 2.3 Fatigue limits of constructional steels versus the parameter C_T (A.V. Bogdanovich, L.A. Sosnovskiy)

From (2.61) it follows that

$$\lg \sigma_{-1T} = \frac{1}{2} \lg C_T; \quad C_T = \left[u_0/\Lambda_{M\backslash T} - a_T T_\Sigma\right] \cdot \frac{1}{a_n} \tag{2.63}$$

According to (2.63), the dependence of limiting stresses on the *parameter of thermomechanic resistance* C_T in the double logarithmic coordinates must be a straight line with the angular coefficient (1/2). The general regularity is as follows: *the higher the value of the parameter* C_T, *the greater is the quantity* σ_{-1T}. Figure 2.3 illustrates a convincing evidence of this dependence for numerous different-grade steels tested for fatigue in different conditions [1, 23, 28]. It is seen that the C_T values vary by more than two orders of magnitude, i.e., by a factor of 100 or more, and the σ_{-1T} values—by more than two orders of magnitude, i.e., by a factor of 10 or more, thus the testing temperature ranged from the helium temperature to $0.8T_s$ (T_S is the melting point). As shown in Fig. 2.3, Eq. (2.63) adequately describes the results of more than 130 experiments.

Equation (2.63) is also analyzed for different-class metal materials using the fatigue test results obtained by many authors and illustrated in Fig. 2.4a. In [43], it is possible to find a list of literature references.

In Fig. 2.4b the similar analysis is made using the tension test results at different temperatures (σ_{uT} is the strength limit). In this case, it is assumed that $\sigma_{-1} = \sigma_{uT}$ in Eq. (2.63). It is obvious: the correlation coefficient is very high—not less than $r = 0.722$ (very occasionally), but in most cases it exceeds $r = 0.9$; the analysis includes more than 300 test results. Works [1, 43] contain other examples of a successful experimental approval of criterion (2.63). This allows us to hope that even more general criteria [for example, Eqs. (2.52), (2.53), (2.60)] will be practically acceptable. In our opinion, further studies must confirm our hope.

As defined above, criterion Eq. (2.60) is valid for $\sigma \leq \sigma_u$. Depending on the test conditions, τ_W can be interpreted as the greatest contact pressure (p_0) in the center of the contact area under rolling. It can also be interpreted as the sliding stress (τ_w) or as the average (nominal) sliding pressure p_a in the contact zone, or as the fretting contact pressure (q). If the value $\sigma = \sigma_{-1}$ is fixed, where $\sigma_{-1} \ll \sigma_u$, then

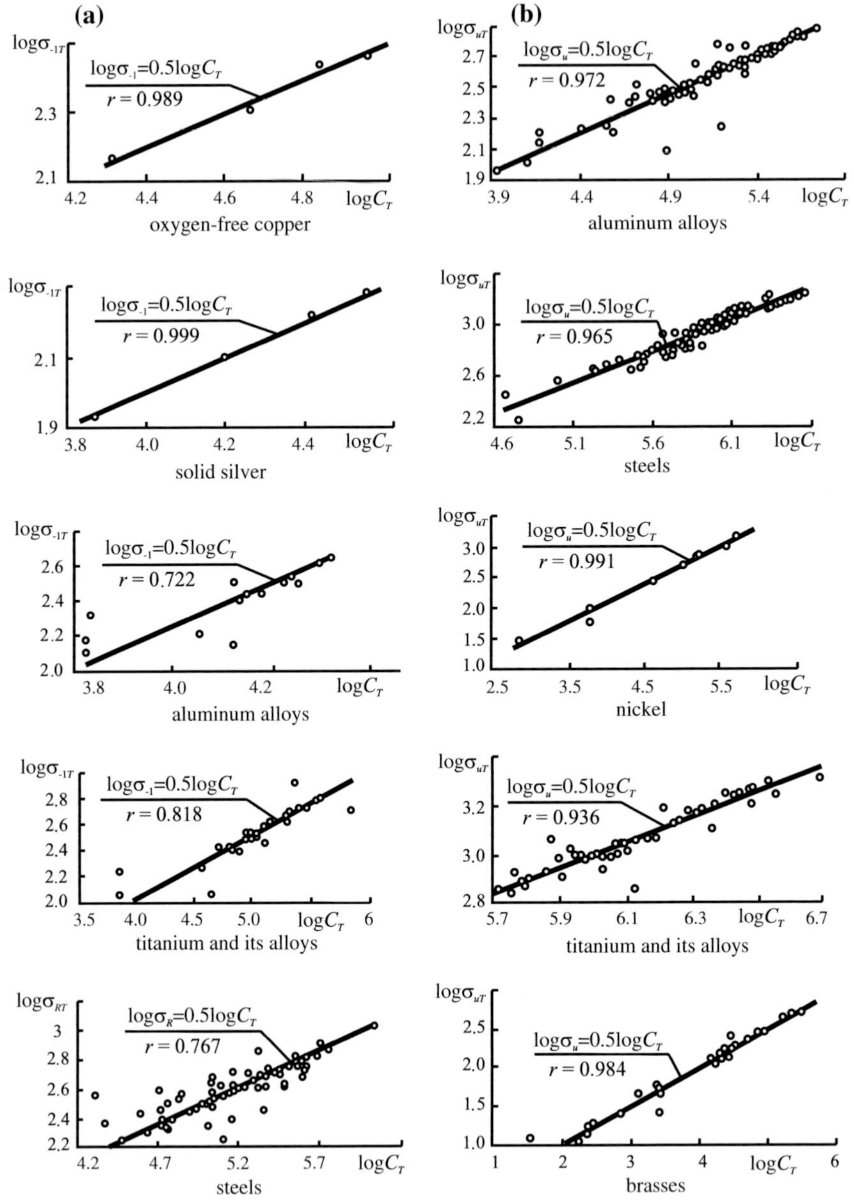

Fig. 2.4 Dependences **a** $\sigma_{-1}(C_T)$ and **b** $\sigma_u(C_T)$ for various metal materials (A.V. Bogdanovich, L.A. Sosnovskiy)

Eq. (2.60) can be represented in the form of the diagram of the limiting states of Tribo-Fatigue systems [1, 28, 37] (Fig. 2.5). The above-mentioned diagram clearly shows the zones of realization of spontaneous hardening-softening processes

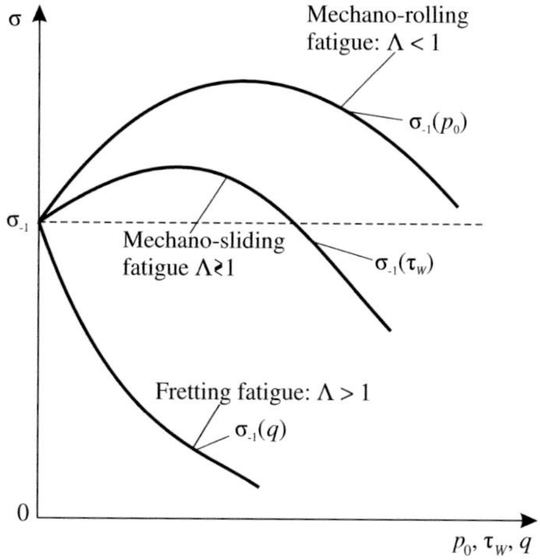

Fig. 2.5 Diagram explaining the basic features of Λ-interactions in the Tribo-Fatigue system

($\Lambda \gtreqless 1$). Figure 2.5 yields the above-mentioned obvious conclusions: if $\Lambda < 1$, then we are dealing with a self-hardening system (during testing or operation under given conditions); if $\Lambda > 1$, then a system turns to be self-softening; if it is found that $\Lambda < 1$ converts to $\Lambda > 1$, then it implies that owing to the changes in the governing conditions of operation or use, the hardening processes are replaced by the softening processes.

Additional experimental support for these conclusions is shown in Figs. 2.6, 2.7 and 2.8. Note that for spontaneous hardening (for $\Lambda < 1$, Figs. 2.5, 2.6 and 2.7), it appears that *the limiting stress in wear-fatigue tests is higher than in routine fatigue tests. It means that in these conditions the processes of friction and wear become "useful"*. There are numerous works (for example, [24]), according to which dosed wear in real Tribo-Fatigue systems (for example, wheel/rail system) causes their fatigue strength to grow. At $\Lambda \gg 1$ (in Fig. 2.8), vice versa, this leads to a strong acceleration of damageability: the fatigue limit decreases with increasing the

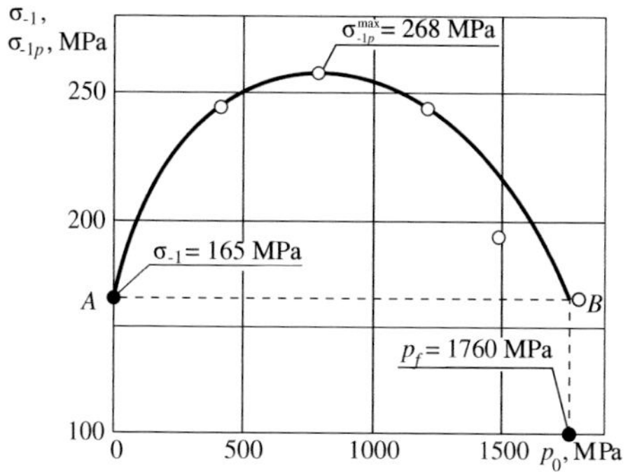

Fig. 2.6 Influence of rolling friction processes on the resistance of mechano-rolling fatigue during tests of the Tribo-Fatigue steel 45 (shaft)/steel 25 HGT (roller) system (L.A. Sosnovskiy, S.A. Tyurin)

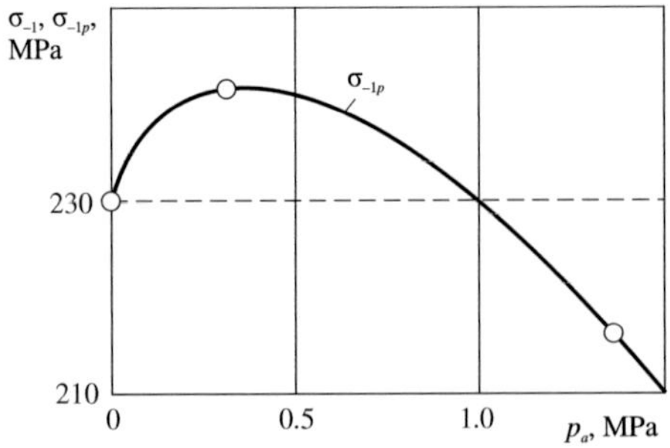

Fig. 2.7 Limiting stresses versus the contact pressure for the Tribo-Fatigue steel 45 (shaft)/cast iron (partial bearing insert) system (V.I. Pokhmursky et al.)

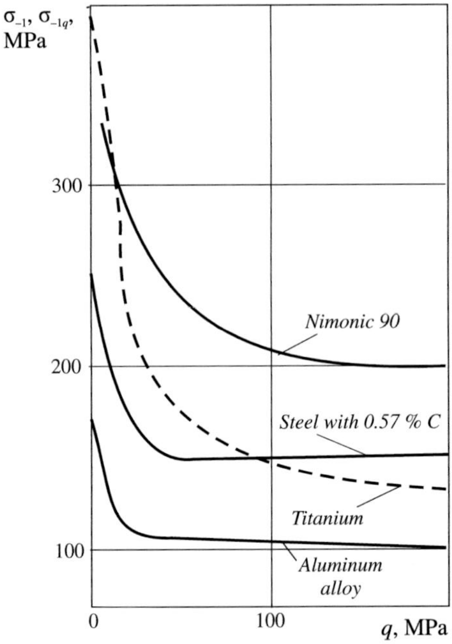

Fig. 2.8 Contact pressure versus changes in the fatigue limit at fretting fatigue according to R.B. Waterhouse (nimonic-90—*Harris W.J.*; steel with 0.5 % C—*Peterson R.E.*; titanium—*Sinclair G. M., Liu H.W., Corten H.T.*; aluminum alloy—*Corten H.T.*)

contact pressure q by a factor of 2–3. In addition, there are many works (for example, [17]), according to which the system wear yields a sharp drop in fatigue strength.

The elements of the theory of Λ-interactions of irreversible damages in active systems are formulated to the present time and are developed to some extent [29–32, 35]. Their physical picture is shown in Figs. 2.9, 2.10 and 2.11 according to

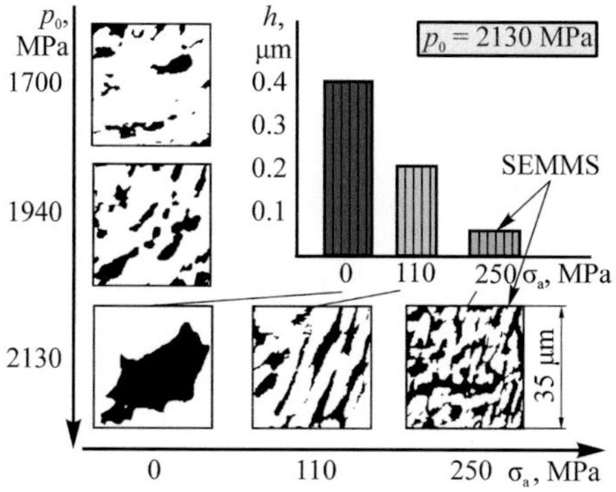

Fig. 2.9 Damage interactions at the submicrolevel: the microtopography of surface damage at rolling friction (*vertical column of pictures*) and mechano-rolling fatigue (*horizontal line of pictures*) (L.A. Sosnovskiy, S.A. Chizhik)

Fig. 2.10 Microlevel: the result **c** of the irreversible interactions of surface damages **a**, **b** due to mechano-rolling fatigue

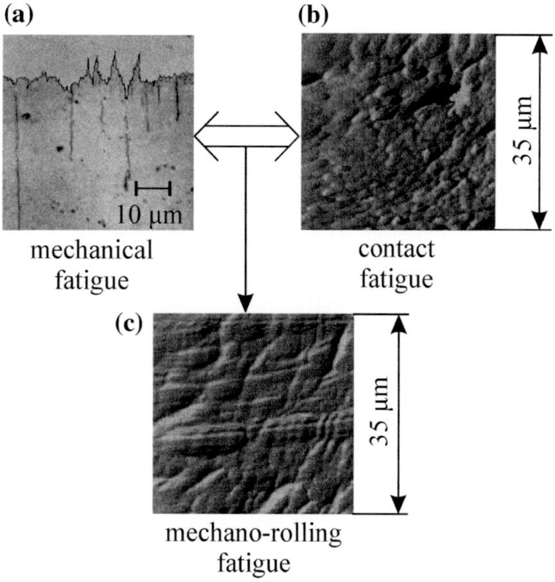

experimental results [3, 26]. As shown in Fig. 2.9, the increase in the contact pressure results in the formation of pitting cavities oriented at roller paths; at p_0 = 2130 MPa their depth reaches 0.4 µm (histogram in the right upper corner). But when in the friction pair the shaft additionally undergoes bending, the damage picture substantially changes even at small cyclic stress values (σ_a = 110 MPa). In what follows, long and deep pitting cavities are not formed. This is the result of interaction of damages caused by different loads—contact and bending.

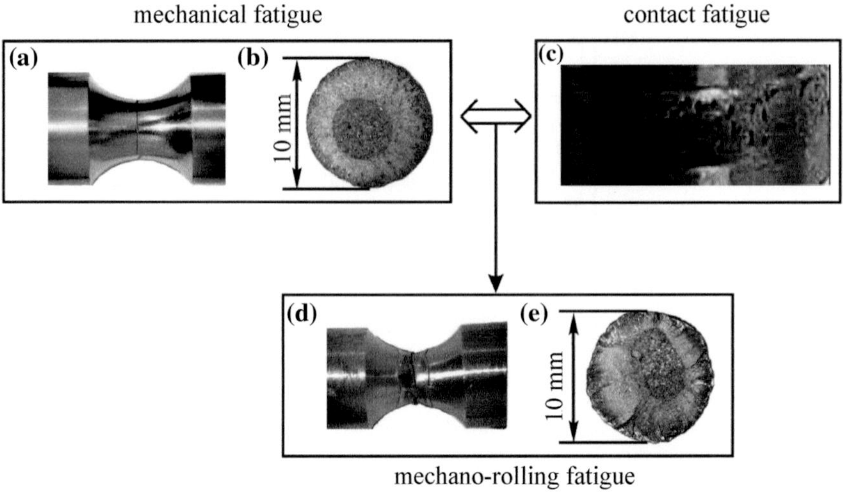

Fig. 2.11 Macrolevel: the result **c**, **d** of the irreversible interactions of shaft damages **a**, **b**, **c** due to mechano-rolling fatigue

Increasing the amplitude of cyclic stresses accelerates the process of formation of the second system of cracks—transverse to the rolling direction. As a result, damage becomes dispersed, there appears almost an equilibrium network of intersecting crack-pores which borders with fine dispersed particles (grain fragments) of material. The higher the cyclic stresses, the smaller and thinner are the separated particles, and the critical depth of a damaged layer decreases to 0.05 μm. Thereby, large and deep pitting cavities are not formed. *Surface crumbling appears to be the dominant process in this case*. It is characterized by the separation of fine dispersed particles from the working surface that are formed due to a multiple shift in intersecting planes, by the formation of a great number of microscopic crack-pores, and by the fine grinding of grains. Such a mechanism of complex surface damage is called the *scattered effect of a multiple microshift* (SEMMS) [34]. It is *Sosnovskiy–Makhutov–Chzhik's effect.*

In Fig. 2.10 it is seen that if at mechanical fatigue over the surface extrusions and intrusions are formed, whereas at contact fatigue the oriented grain topography of surface damages is formed, then in the case of mechano-rolling fatigue (due to corresponding damage interactions) another type of degradation is formed: the broken picture of multiple microshifts intersecting in two planes.

Integrally similar interactions also form the fundamentally different microstructure of fracture (Fig. 2.11).

Tables 2.1 and 2.2 summarize the physical signs of different (often encountered in practice) features of the limiting state that can find use in the corresponding research areas.

Table 2.1 Main physical signs of the limiting state

State		Condition of reaching the limiting (critical) state
Symbol	Physical state and its characteristic	
M	Mechanical state σ_{ij}	$u_n^{eff} \xrightarrow{\sigma_{ij} \to \sigma_{\lim}} u_0$
T	Thermodynamic state T_Σ	$u_T^{\text{eff}} \xrightarrow{T_\Sigma \to T_S} u_0$
MTD	Mechanothermodynamic state σ_{ijT}, T_Σ	$u_\Sigma^{eff} \underset{T_\Sigma \to T_S}{\overset{\sigma_{ijT} \to \sigma_{\lim}(T)}{\Longrightarrow}} u_0$
tMTD	Mechanothermodynamic state in time $\sigma_{ijT}, T_\Sigma, t$	$u_\Sigma^{eff} \underset{\substack{T_\Sigma \to T_S \\ t \to t_{\lim}}}{\overset{\sigma_{ijT} \to \sigma_{\lim}(T)}{\Longrightarrow}} u_0$

Here: $\sigma_{\lim}$ is the limiting stress; T_S is the melting point; $t_{\lim}$ is the longevity; σ_{ij} is the stress (strain) tensor; T_Σ is the temperature due to all heat sources; σ_{ijT} is the stress tensor in the isothermal ($T_\Sigma = \text{const}$) state; σ_{ijT}, T_Σ is the stress-strain state and the thermodynamic state, respectively; $\sigma_{ijT}, T_\Sigma, t$ is the stress-strain state and the thermodynamic state in time, respectively

Table 2.2 Concrete definition of characteristics and their physical signs of the limiting state

Criterion condition	Condition of reaching the limiting state	Physical sign
L1	$\sigma_{\lim} = \sigma_u$	Static failure
	σ_u—tensile strength limit	
L2	$\sigma_{\lim} = \sigma_{-1}$	Fatigue failure (into parts)
	σ_{-1}—mechanical fatigue limit	
L3	$\sigma_{\lim} = p_f$	Pitting cavities of critical density (critical depth), excessive wear
	p_f—rolling fatigue limit	
L4	$\sigma_{\lim} = \tau_f$	Limiting wear
	τ_f—sliding fatigue limit	
L5	$\sigma_{\lim} = \begin{cases} \sigma_{-1p} \\ \sigma_{-1\tau} \end{cases}$	Fatigue failure (into parts) depending on contact pressure (subscript p) at rolling or on friction stress (subscript τ) at sliding (direct effect in Tribo-Fatigue)
	$\sigma_{-1p}, \sigma_{-1\tau}$—stress limit during the direct effect implementation	
L6	$\sigma_{\lim} = \begin{cases} p_{f\sigma} \\ \tau_{f\sigma} \end{cases}$	Pitting cavities of critical density (critical depth) or excessive wear at rolling or sliding depending on the level of cyclic stresses σ (subscript σ) (back effect in Tribo-Fatigue)
	$p_{f\sigma}, \tau_{f\sigma}$—stress limit during the back effect implementation	
L7	$\sigma_{\lim} = \sigma_{-1q}$	Fatigue failure at fretting corrosion and (or) fretting wear
	σ_{-1q}—fretting fatigue limit	
L8	$\sigma_{\lim T} = \sigma_{-1T}$	Limiting state depending on temperature (isothermal fatigue)
	σ_{-1T}—isothermal fatigue limit	
L9	$T_{\lim} = T_S$	Thermal (thermodynamic) damage
	T_S—melting point	
L10	$t_{\lim} = t_c$	Time (physical) prior to the onset of the limiting state on the basis of any sign
	t_c—longevity	

As for the determination of the parameters $\Lambda_{M\backslash T}$ and $\Lambda_{n\backslash\tau}$, it is shown in [1, 37, 43] that, for example, the parameter $\Lambda_{n\backslash\tau}$ is the function of the *relative (with respect to the limiting state) skewness coefficient* [see (2.45)] *of wear-fatigue damage:*

$$\bar{\rho}_{n\backslash\tau} = \left(\frac{\tau_w}{\tau_f}\right)^2 \left(\frac{\sigma_{-1}}{\sigma}\right)^2. \tag{2.64}$$

Hence, it follows that $\bar{\rho}_{n\backslash\tau}$ depends not only on the absolute values of effective (σ, τ_w) and limiting (σ_{-1}, τ_f) stresses, but also on their ratios, namely: $\tau_w/\sigma, \sigma_{-1}/\tau_f$, σ_{-1}/σ, $\tau_w/\tau_f \gtreqless 1$. This means, for example, that significantly different patterns of irreversible damage accumulation will be implemented depending on the realization of this or that of the inequalities $\sigma \gtreqless \sigma_{-1}$, $\tau_w \gtreqless \tau_f$. This conclusion corresponds to the known experimental results and theoretical models. Figure 2.12 shows the analysis with regard to the possible dependences $\log \Lambda_{n\backslash\tau} - \log\bar{\rho}_{n\backslash\tau}$ [37, 43]. A more detailed analysis of the interdependences $\Lambda_{n\backslash\tau}\left(\bar{\rho}_{n\backslash\tau}\right)$ can be found in [1, 37, 43].

The dependence of the interaction parameter $\Lambda_{T\backslash M}$ on the parameter $\bar{\rho}_{T\backslash M}$ can be analyzed in a similar way. Such a dependence of steel, aluminum alloys, and nickel (according to the extensive experimental results [1, 37, 43]) in the double logarithmic coordinates is shown in Fig. 2.13. The correlation coefficient has appeared to be very high: from $r = 0.862$ to $r = 0.999$. The dependence $\Lambda_{T\backslash M}(\bar{\rho}_{T\backslash M})$ as a rule, undergoes sudden changes for $\lg\ \bar{\rho}_{T\backslash M} = 0$, i.e., at the value $\bar{\rho}_{T\backslash M} = 1$ when thermal and force damages appear to be equilibrium (as compared to the similar changes in the plots in Fig. 2.12).

For steels and nickel, at $\bar{\rho}_{T\backslash M} < 1$ the direct dependence is found between $\Lambda_{T\backslash M}$ and $\bar{\rho}_{T\backslash M}$, and at $\bar{\rho}_{T\backslash M} > 1$ it becomes inverse. For aluminum alloys, the dependence $\Lambda_{T\backslash M}(\bar{\rho}_{T\backslash M})$ is also direct, but it is located (at $\bar{\rho}_{T\backslash M} < 1$) in III quadrant.

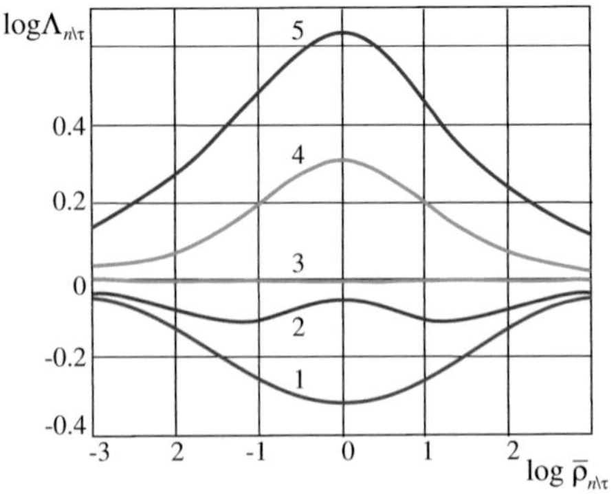

Fig. 2.12 Typical character and direction of hardening-softening processes ($\Lambda \gtreqless 1$) versus the skewness coefficient of damageability $\bar{\rho}$: *1*, *2*—mechano-rolling fatigue; *2*, *3*, *4*—mechano-sliding fatigue; *4*, *5*—fretting fatigue

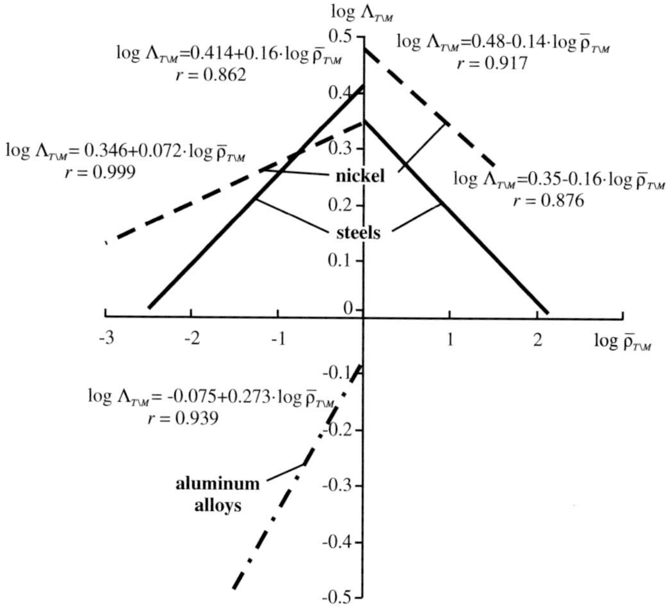

Fig. 2.13 Logarithmic dependences of $\Lambda_{T\backslash M}(\bar{\rho}_{T\backslash M})$ built using the experimental data (L.A. Sosnovskiy, A.V. Bogdanovich)

Thus, it is experimentally confirmed that *the interaction parameter* $\Lambda_{T\backslash M}$ *is sensitive not only to the effective thermal-to-mechanical energy ratio, but also to the structure and composition (or nature) of metal materials*. The last conclusion is also valid for the parameter $\Lambda_{n\backslash\tau}$: its numerical values appear to be significantly different, for example, for metal/metal and metal/polymer active systems—even in the case when the ratios $\sigma\backslash\sigma_{-1}$ and $\tau_w\backslash\tau_f$ are identical for them.

2.7 Summary of Experimental Data

The data of more than 600 tests of metals and their alloys (under isothermal conditions) obtained by many authors are briefly analyzed and have been presented above (see Sect. 2.6). It was found that the thermodynamic dependence of limiting stresses can be represented in the $\log\sigma_{\lim} - \log C_T$ coordinates (Figs. 2.3 and 2.4 and formula (2.63), where the function

$$C_T = C_T(T, u_0, a_n, a_T, \Lambda_{M\backslash T}) \tag{2.65}$$

is satisfactory under both the conditions of static tension $(\sigma_{\lim} = \sigma_u)$ and fatigue failure $(\sigma_{\lim} = \sigma_{-1})$ for numerous and different metal materials (steels; aluminum,

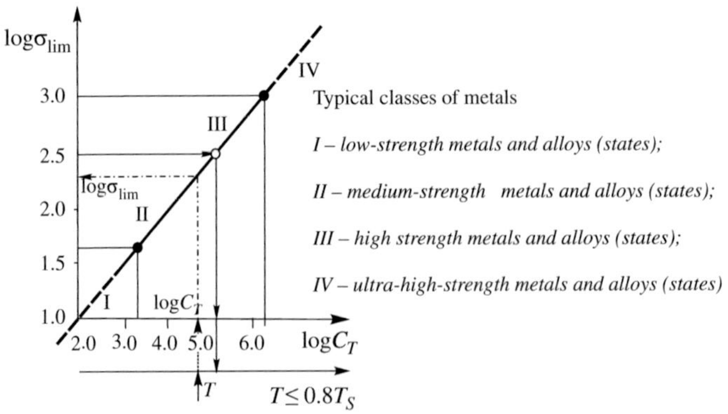

Fig. 2.14 Generalized MTD function of limiting states of metals and alloys $\sigma_{\lim} \leq \sigma_u$; $T_\Sigma \leq 0.8T_S$

titanium, and other alloys, etc.). In addition, interrelation (2.63) appears to be valid practically within the entire possible ranges of temperature $(T_\Sigma \leq 0.8T_S)$ and stress $(\sigma \leq \sigma_u)$ with the correlation $r = 0.7$ in the specific cases and usually with $r > 0.9$. Model (2.63) then turns to be *fundamental* (Fig. 2.14). First, this simplified model might seem to be questionable since in the known literature [4, 42, etc.], the explicit temperature dependence of limiting stresses is described by means of complex curves. This is attributed to the changes in the mechanisms of damage of different materials under various testing conditions—at normal, operating, and other temperatures. Nevertheless the fundamental nature of model (2.63) is convincingly confirmed experimentally (Figs. 2.3 and 2.4).

From the theoretical standpoint, the following considerations speak in favor of model (2.63). It has four parameters [formula (2.65)], and one of them (u_0) is a fundamental constant of substance [formulas (2.24), (2.25)], and two others (a_T, a_n) are defined by boundary conditions (2.54) as the relations for u_0 and physical constants σ_d and T_d of a given material:

$$a_n = u_0/\,\sigma_d^2, a_T = u_0/T_d. \tag{2.66}$$

The methods of σ_d and T_d determination are described in [1, 37, 43]. Here we remind that the material destruction limit σ_d is determined under the tension conditions as $T_\Sigma \to 0$ and the material destruction temperature T_d—at the solid heating for $\sigma = 0$. As can be seen from the above, the *dual character of accumulation processes of damage and failure* caused by (1) mechanical stress and (2) thermal activation of this stress in time [48] is considered in the general case. Finally, as briefly described above and outlined in [18, 37], the function $\Lambda_{M\backslash T} \gtreqless 1$ considers the interaction of damages due to changes in $\sigma \gtreqless \sigma_{\lim}$. In the known studies (see, for example [14]), it is also convincingly proved many times that just this relation is responsible for the character and damage mechanisms at elastic, inelastic, elasto-plastic and plastic strain. Also, the role of thermal fluctuations ($T_{\Sigma} < T_d$) is, for example, studied in detail in [16, 48].

It remains for us to put the "last point" in the argument in favor of the fundamental character of model (2.63). If it is really fundamental, then it must also be valid for non-metal, for example, polymer materials—according to hypothesis (2.24). The analysis results of polymer tests based on experimental data are presented in Fig. 2.15 and in Table 2.3. It is seen that model (2.63) is verified by the correlation coefficient $r = 0.917$. Note that the test results for not only "normal" samples (with a diameter of ~ 5 mm), but also for thin polymer threads and films

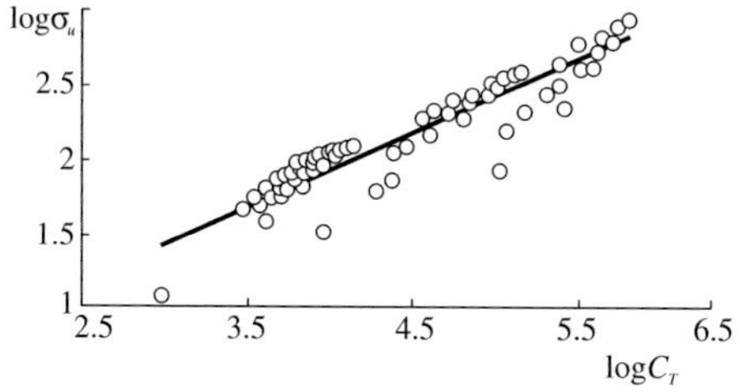

Fig. 2.15 Dependence $\sigma_u(C_T)$ for polymer materials (A.V. Bogdanovich)

Table 2.3 Main characteristics of polymer materials analyzed in terms of the energy criterion [15, 21]

Material and reference	$u_0, \frac{\text{kJ}}{\text{mol}}$	$\frac{a_T}{a_n}, \frac{\text{MPa}^2}{\text{K}}$ $\left(\frac{\text{kJ}}{\text{mol K}} / \frac{\text{kJ}}{\text{mol MPa}^2}\right)$	Test data	
			$\frac{\text{K}}{\sigma_b, \text{MPa}}$	Sample size
High-density polyethylene film (HDPF), grade 20806-024	108	$\frac{0.275}{2.94\times10^{-4}}$	$\frac{275-383}{32-386}$	5
Polypropylene film (PF), grade 03P10/005	119	$\frac{0.234}{1.70\times10^{-4}}$	$\frac{273-423}{150-570}$	5
Hardened staple fiber made of polyvinyl alcohol (PVA) "Vinol MF"	111	$\frac{0.227}{7.62\times10^{-5}}$	$\frac{273-453}{80-802}$	5
Thread based on perchlorvinyl resin (PCV), grade "Chlorine"	114	$\frac{0.285}{2.56\times10^{-4}}$	$\frac{273-383}{60-376}$	5
Caprone thread, (GOST 7054067)	169	$\frac{0.282}{1.68\times10^{-4}}$	$\frac{275-453}{300-740}$	5
Polyethylene terephthalate film (PET), (TU 6-05-1597-72)	222	$\frac{0.342}{9.82\times10^{-4}}$	$\frac{279-498}{200-362}$	4
Polyamide film PM-1, (TU 6-05-1597-72)	202	$\frac{0.297}{2.1\times10^{-3}}$	$\frac{273-673}{12-240}$	7
Polystyrol (PS) at bending	281	$\frac{0.627}{2\times10^{-2}}$	$\frac{77-290}{56-108}$	10
Polymetalmethacrylate (PMMA) at bending	277	$\frac{0.558}{1.74\times10^{-2}}$	$\frac{77-290}{66-116}$	10
High-impact polystyrene (HIPS) at tension and torsion	277	$\frac{0.699}{2.53\times10^{-2}}$	$\frac{77-290}{48-94}$	10
	252	$\frac{0.636}{1.84\times10^{-2}}$	$\frac{77-290}{50-105}$	10

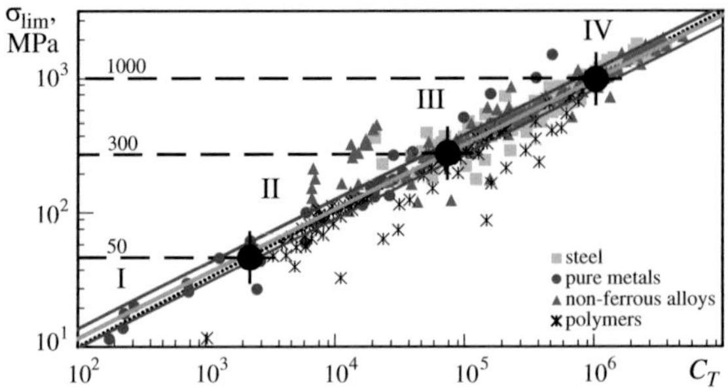

Fig. 2.16 Experimentally justified MTD function of states critical with respect to damageability for pure metals, non-ferrous alloys, structural steel and polymer materials

are processed not only at tension, but also at torsion and bending. A large deviation of several points from the fundamental straight line is due to the fact that for these results $\Lambda_{M\backslash T} = 1$ is conventionally assumed thanks to the lack of experimental data in effort to assess the real value of this parameter.

The *generalized experimentally justified MTD function of the limiting states (*in terms of *damageability)* is shown in Fig. 2.16. Relatively large deviations of particular experimental points from the predicted ones are also seen in Figs. 2.3 and 2.4 for two reasons: either available references lack sufficient data for a correct assessment of required parameters, or the conducted experiments contain significant errors or they are not quite correct methodically. A possible analysis of other researchers will show this was true or not.

Note that model (2.63) may seem to be non-fundamental because of its simplicity. However, remind the saying that has become a classic dictum: the fundamental dependence cannot be complicated (or: any law is described by the simplest formula. Thus, model (2.63) can serve for prediction (shown by the arrows from T to $\sigma_{\lim}$ in Fig. 2.14) of the *mechanical behavior of materials in the thermodynamic medium*:

$$T \xrightarrow[\substack{\uparrow \\ a_n,\, a_T}]{\substack{u_0,\, \Lambda_{M\backslash T} \\ \downarrow}} \lg C_T \rightarrow \lg \sigma_{\lim}\left(T, u_0, a_n, a_T, \Lambda_{M\backslash T}\right) \rightarrow \sigma_{\lim(T)}. \tag{2.67}$$

The state of the medium in (2.67) is described with the use of the parameters T, a_T and $\Lambda_{M\backslash T}$.

As seen, the predictions by (2.63) and (2.67) are applicable for the materials of different nature and structure—irrespective of damage and failure mechanisms at static and cyclic loads. It would be interesting to make a similar analysis of the tests at impact, but such an analysis lies outside the scope of the present work.

Certainly, due to the linearity of function (2.63), the *reverse prediction* appears to be possible and effective. If it is necessary to have a given mechanical state of material (determined by u_0, $\sigma_{\lim(T)}$), then the requirements can be formulated to the

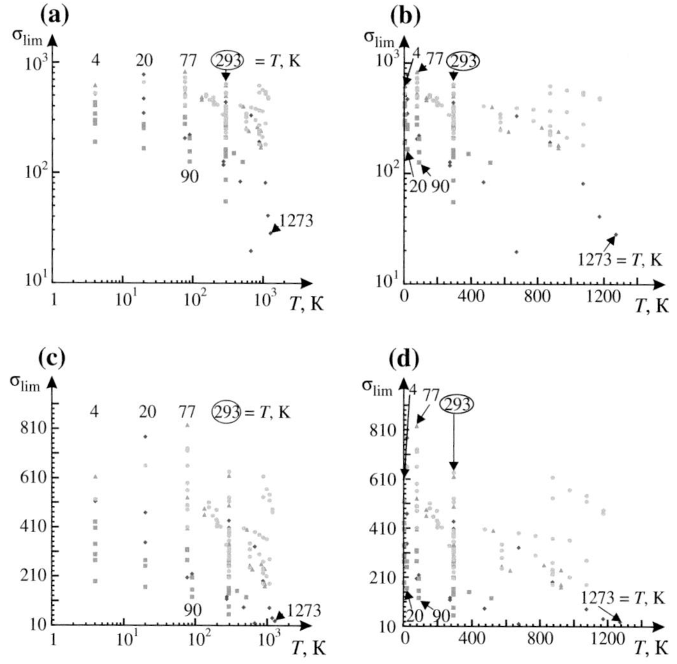

Fig. 2.17 (Beginning) Explicit temperature dependences of the fatigue limit for structural steels in logarithmic **a**, semi-logarithmic **b**, **c** and uniform **d** coordinates

medium (determined by the parameters T, a_T, $\Lambda_{M\backslash T}$) in which the system can operate (shown by the arrows from σ_{lim} to T in Fig. 2.14):

$$\sigma_{\text{lim(T)}} \to \lg \sigma_{\text{lim(T)}} \xrightarrow[\substack{\uparrow \\ a_n,\, a_T}]{\substack{u_0,\, \Lambda_{M\backslash T} \\ \downarrow}} \lg C_T \to C_T\big(T, u_0, a_n, a_\tau, \Lambda_{M\backslash T}\big) \to T. \tag{2.68}$$

Note that the attempts to construct the explicit temperature dependence of limiting stresses in uniform, semi-logarithmic and logarithmic coordinates for different materials and various test conditions are quite ineffective because a relatively small amount (136) of test results, as shown in Fig. 2.17.

Further, we briefly analyze a more complex *problem of the MTD system operation in the medium* in which the processes of *thermal corrosion and stress corrosion* are implemented. From (2.52), at $\tau_w = 0$ we have

$$\Lambda_{M\backslash T}\left(\frac{a_T}{1 - D_T} T_\Sigma + \frac{a_n}{1 - D_n}\sigma^2\right) = u_0. \tag{2.69}$$

Upon simple manipulations, we obtain

$$\sigma_{\lim(T,ch)} = \frac{1}{2}\lg C_{T(ch)} \tag{2.70}$$

where, as can be easily shown, the parameter of resistance to thermal stress corrosion is:

$$C_{T(ch)} = C_{T(ch)}\left(T, u_0, a_n, a_T, \Lambda_{M\backslash T}, v_{ch}, v_{ch(\sigma)}, m_{v(\sigma)}, v_{ch(T)}, m_{v(T)}\right). \tag{2.71}$$

It can be seen that laws (2.63) and (2.70) are fundamentally (and formally) identical and differ in the fact that appropriate functions (2.65) and (2.71) take account of those parameters which describe the damageability processes characteristic for the phenomena analyzed. So, in (2.71) the parameters $v_{ch}, v_{ch(\sigma)}, m_{v(\sigma)}, v_{ch(T)}, m_{v(T)}$ describe the processes of thermal stress corrosion. Based on (2.70) and (2.71), it is easy to build *prediction algorithms* [of form (2.67) and (2.68)] *of resistance to thermal stress corrosion*.

A further and detailed analysis of (2.70) and (2.71) is beyond the scope of this study.

Note that solutions (2.52)–(2.62) can be analyzed similarly for other testing (or operating) conditions.

Thus, a single MTD function of critical with respect to damageability states of metals and polymers working in various conditions has been obtained above. The analysis of more than 600 experimental results (Figs. 2.3, 2.4, 2.13, 2.15 and 2.16) showed that this function is fundamental: it is valid for low-, mean- and high-strength states of pure metals, alloys, and polymers over a wide range of medium temperatures (from helium to $0.8T_S$, where T_S is the melting point of material) and mechanical loads (up to the strength limit for single static loading); the fatigue life was of the order of 10^6–10^8 cycles.

The fundamental MTD function as found in the present study can be used for effective prediction of the behavior of particular MTD systems in various operating (test) conditions. Model (2.70), (2.71) is proposed for the description of the influence of thermal corrosion and stress corrosion on the changes in the limiting states of materials.

2.8 Translimiting States

According to the available information, the *theory of translimiting states* is not yet sufficiently developed [37]. The elements of this theory will be set forth on the basis of solutions (2.47), (2.51) and (2.52).

Figure 2.18 shows the general analysis of the contribution of *mechano-chemico-thermal damage* (parameters *D*) to the process of reaching the limiting state by the

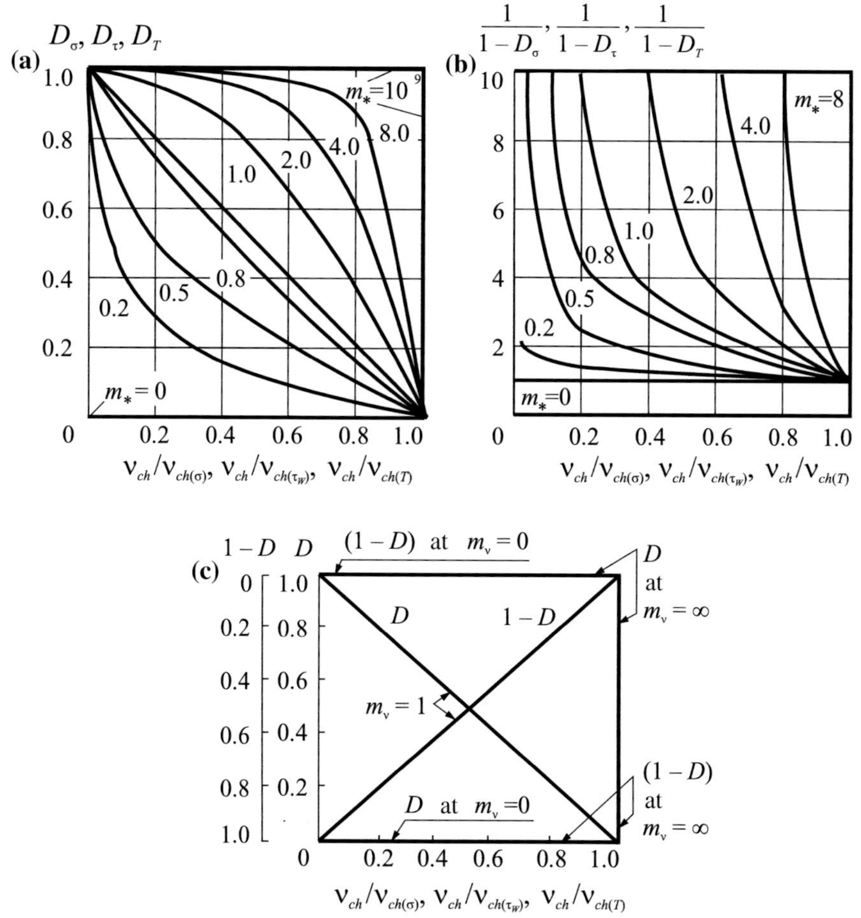

Fig. 2.18 Analysis of the influence of mechano-chemico-thermal processes on the system damageability

MTD system. Based on the analysis of formulas (2.47)–(2.52) and Fig. 2.18, the following conclusions can be drawn.

1. The growth of parameters D means a decrease in a relative damage rate $v_{ch}/v_{ch(*)}$ (Fig. 2.18a). In other words, mechano-chemico-thermal damage accelerates the achievement of the limiting state by the MTD system the faster, the greater is the value of the parameter D and/or of the rate $v_{ch(*)}$.
2. The parameter m_v exerts the strongest influence on the system damage, and it is the stronger, the larger is its value (Fig. 2.18b). The important feature of this influence is that this environment is very sensitive to the excitation of mechanical stresses in the MTD system and to the temperature rise if for it the parameter $m_v > 5$. In other words, in such a case, the translimiting state can be realized, for which damageability measure (2.29) is more than unity ($\psi_u^{eff} > 1$),

whereas according to (2.28), it is sufficient to have $\psi_u^{eff} = 1$ to reach the limiting state

Two specific cases are illustrated in Fig. 2.18c.

(1) $D = 0$. There is no electrochemical corrosion influence on wear-fatigue damage. But this doesn't mean that the electrochemical corrosion process does not occur. In fact, according to (2.51), when $D = 0$, we have (if $m_v = 1$):

$$1 - \frac{v_{ch}}{v_{ch(*)}} b_* = 0.$$

This implies that the situation should be the following: $b_* = 1$ and $v_{ch} / v_{ch(*)} = 1$, i.e., the corrosion rate is insensitive to this factor (mechanical or frictional stress). This means that *threshold values* of σ^0, τ_w^0, and T^0 exist for a given environment. The corrosion rate in such an environment does not vary for $\sigma \leq \sigma^0$, $\tau_w \leq \tau_w^0$ and for $T_\Sigma \leq T^0$ [formula (2.52)].

(2) $D = 1$ and, hence, $1/(1 - D) \to \infty$. Explosive damage is realized within the system as $\psi_u^{eff} \to \infty$. In this case, it should be

$$\frac{v_{ch}}{v_{ch(*)}} b_* = 0.$$

Since $v_{ch} = 0$ is impossible, it can be assumed that $v_{ch(*)} \to \infty$. This is just the *condition of mechano-chemico-thermal explosion* in the MTD system. The explosion is not just due to the environmental influence—the environmental influence dramatically enhanced by temperatures and mechanical stresses.

Thus, complex function (2.47) for the damageability of MTD systems can also be used for *analyzing their translimiting states* caused by a supercritical growth of thermodynamic, mechanical, frictional, and electrochemical loads according to formulas (2.48)–(2.51), i.e.,

$$1 \leq \psi_u^{eff} = \Lambda_{T\backslash M} \left[\psi_{T(ch)} + \Lambda_{n\backslash\tau} (\psi_{n(ch)} + \psi_{\tau(ch)}) \right] \leq \infty. \quad (2.72)$$

According to (2.72), there are *many translimiting states* of the MTD system defined by the condition $\psi_u^{eff} > 1$. This is possible in those (many) cases when the critical with respect to damageability state of the system is reached not at one but at many points of the dangerous volume. Hence the assumption can be made that *many (different) forms* of these states must exist. As an example, some forms of translimiting states of a real wheel/rail system observed in operating and testing conditions [36, 37] are presented in Figs. 2.19 and 2.20.

In [22], it is possible to find the form of function (2.72) and its analysis for the simplest Tribo-Fatigue system (shaft/sliding bearing).

Although above-mentioned criterion Eqs. (2.19), (2.23), (2.28), (2.33), (2.47), (2.52) are obtained from the consideration of the energy conditions of reaching the limiting state, it is stated that they can in principle be used for describing a variety

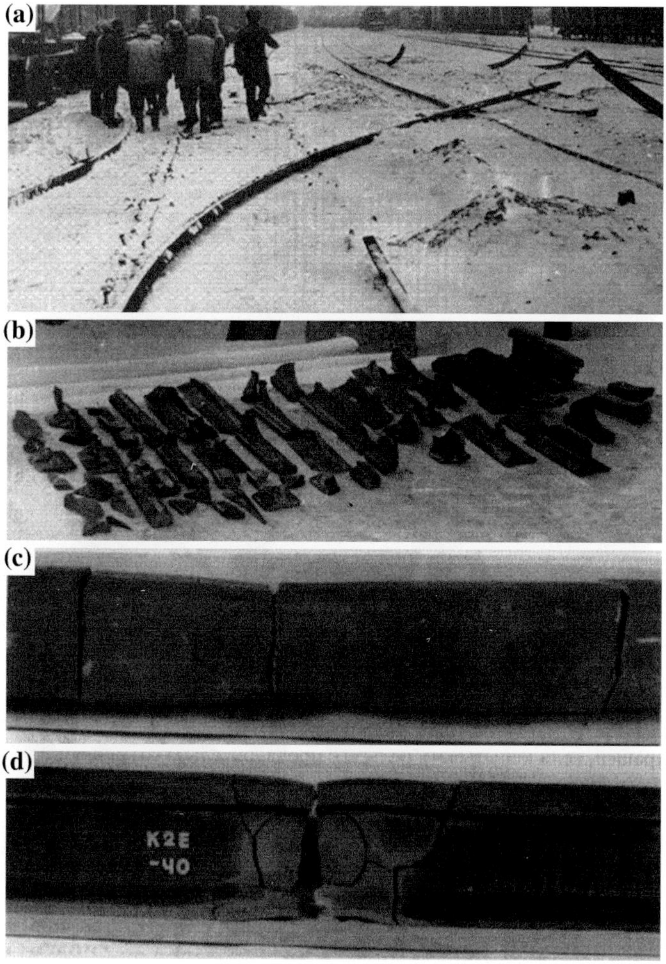

Fig. 2.19 Four forms of the translimiting state of rails appeared in the operation **a**, **b** (V.I. Matvetsov), and detected in the three-point bending tests **c**, **d** (M.N. Georgiev et al.)

of translimiting states, but only in those cases when situations in the MTD system are created for an *unconditional supercritical* (essentially *unrestrained*) growth of loads (explosions, accidents, disasters, fires, etc.).

Another more general approach for the analysis of translimiting states is that it considers a *damage space* defined according to (2.34), (2.39) by volumetric measures

$$0 \leq \omega_{ij} = \frac{V_{ij}}{V_0} \leq 1. \tag{2.73}$$

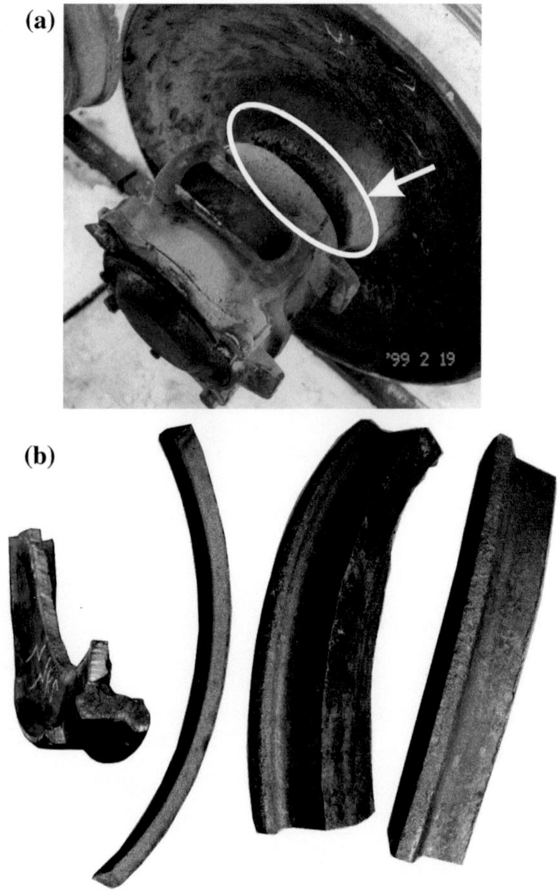

Fig. 2.20 Two forms **a**, **b** of the translimiting state of the freight car wheel appeared in the operation (I.F. Pastukhov)

On the basis of (2.47)–(2.51), spatial measures of damageability can be defined as

$$\omega_{n(ch)} = \frac{V_{P_\gamma}}{V_0(1 - D_n)}, \ \omega_{\tau(ch)} = \frac{S_{P_\gamma}}{S_0(1 - D_\tau)}, \ \omega_{T(ch)} = \frac{V_{T_\gamma}}{V_0(1 - D_T)}, \tag{2.74}$$

where V_0, S_k are the working volume and the surface, respectively. So criterion (2.33) can be written with regard to (2.74):

$$\Lambda_{T\backslash M}\left[\frac{V_{T_\gamma}}{V_0(1 - D_T)} + \Lambda_{n\backslash\tau}\left(\frac{V_{P_\gamma}}{V_0(1 - D_n)} + \frac{S_{P_\gamma}}{S_0(1 - D_\tau)}\right)\right] = 1. \tag{2.75}$$

The advantage of (2.75) is that the interaction of dangerous volumes [37] at different loads is taken into account when the limiting state of MTD systems is formed. In addition, as mentioned above, since absolute dangerous volumes are determined by a number of structural-technological and metallurgical factors (2.39), these factors appear to be automatically accounted for in the limiting state criterion for such systems.

If the rupture of atomic bonds is realized only over one dangerous section of an object at all "points" of this section ($u_{\Sigma}^{eff} = u_0$), then it is divided into two parts, which corresponds to the condition $\omega_{\Sigma} = 1$. But if the complex of loads (mechanical, electrochemical, thermodynamic, etc.) is such that the rupture of "all" atomic bonds takes place over this section, then the process occurs and it is called *disintegration of an object,* whose death corresponds to the condition $\omega_{\Sigma}^{*} = \infty$. This is the most *common form of the translimiting state*: the system disintegrates into an infinite number of particles of arbitrarily small size (for example, atoms). It is clear that there must be some *intermediate forms of the translimiting states of the system.* The condition of their implementation is

$$1 \leq \omega_{\Sigma}^{*} = \Lambda_{T\backslash M}\left[\left(\omega_{n(ch)} + \omega_{\tau(ch)}\right)\Lambda_{\sigma\backslash\tau} + \omega_{T(ch)}\right] \leq \infty. \quad (2.76)$$

Naturally, Eq. (2.76) is similar to (2.72). Their difference is that conditions (2.72) are written in terms of energy measures of damage, while conditions (2.76)—in terms of volumetric (spatial) measures of damage.

The general *classification* of conceivable *states of an object* in terms of *volumetric damage* is given in Table 2.4 that is similar to the one in Table 2.1 [25], but with the difference that a special index (asterisk*) is introduced for translimiting states.

The probability interpretation [33, 44] of irreversible damage events in the MTD system can be made according to Table 2.4 and condition (2.76).

If

$$0 \leq P(\omega_{\Sigma}) \leq 1 \quad (2.77)$$

is the *classical probability of the MTD system failure* in terms of *damageability* $(0 \leq \psi_{\Sigma} \leq 1)$ within the time interval $(t_0, T_{\oplus})$ (Item XIV), then $P(\omega_{\Sigma} = \omega_c = 1) = 1$ is the *reliable probability of unconditional functional failure*. For transmitting states the *concept of reliable probability* [33] is introduced

$$1 \leq P_{*}\left(\omega_{\Sigma}^{*}\right) \leq \infty. \quad (2.78)$$

Table 2.4 Characteristic of the states of objects

A-state	Undamaged	$\omega_{\Sigma} = 0 = \omega_0$	↓ *A*-evolution: characteristic system states in terms of damageability
B-state	Damaged	$0 < \omega_{\Sigma} < 1$	
C-state	Critical (limiting)	$\omega_{\Sigma} = 1 = \omega_c$	
D-state	Supercritical (translimiting)	$1 < \omega_{\Sigma}^{*} < \infty$	
E-state	Disintegration	$\omega_{\Sigma}^{*} = \infty = \omega_{\infty}^{*}$	

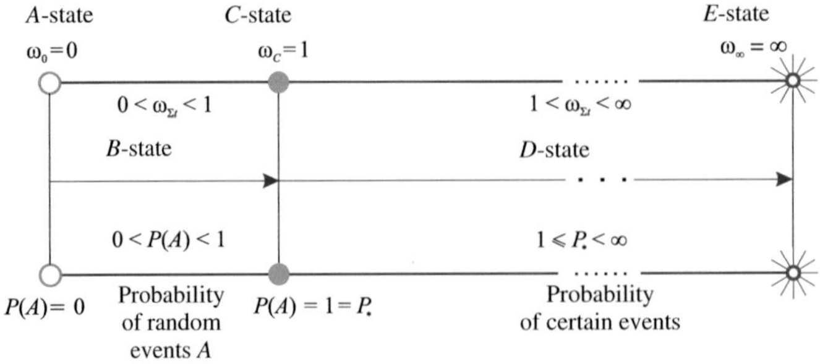

Fig. 2.21 Relationship between the system damages and the probability

These *supercritical damages* $1 < \omega_{\Sigma}^{*} < \infty$ correspond to numerous and infinite shapes and sizes of fragments or particles to be formed in the process of degradation (disintegration) of the system.

Figure 2.21 illustrates the relationship between the system damages and the probability.

Note that the data in Table 2.4 can be interpreted in the following way. If similar to (2.7) the damageability is

$$\omega_{\Sigma}^{*} \to \infty, \tag{2.79}$$

then the absolute size of forming particles should be as small as desired according to (2.8), i.e.,

$$d_{\omega}^{*} \to 0. \tag{2.80}$$

Assume the logarithmic relationship between d_{ω} and ω_{Σ} to a first approximation. The law of degradation

$$d_{\omega}^{*} = e^{-\omega_{\Sigma}^{*}} \text{ or } \omega_{\Sigma}^{*} = -\ln d_{\omega}^{*}. \tag{2.81}$$

As follows from the above-mentioned, all states of the MTD system are predicted by appropriate Eq. (2.72) and/or (2.76). A drawback of this prediction or the description is that the dependence of damageability measures [for example, (2.72)] on the determining parameters appears to be smooth over the entire range $0 \leq \omega_{\Sigma} \leq \infty$ (Fig. 2.22a). It should be noted, however that this is valid only in the case (essentially, in the ideal case) when the values of the determining parameters (σ, τ_w, Λ, etc.) are continuously increasing. But the *surface of damageability reveals jumps (discontinuities)* whenever either discontinuities of any load or abrupt changes in hardening-softening processes (Fig. 2.22b, c) are realized. It is easy to understand that in reality, these specific situations lead to *damageability discontinuities*, i.e., to *qualitative changes* or *system state transformations.* It should be added that our approach has a special advantage: it is based on the analysis of *damageability as a physical reality* independent of the fact what damage mechanisms are already known to us and what mechanisms will be clarified.

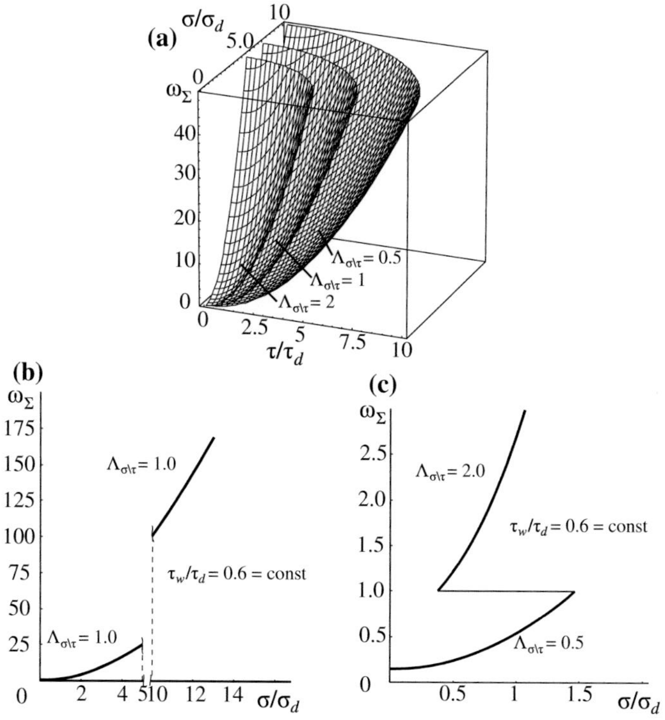

Fig. 2.22 Formation of damageability surfaces **a** and functions **b**, **c** ω_Σ due to the changes in the determining parameters ($\sigma / \sigma_d > 0$, $\tau / \tau_d > 0$, $\Lambda_{\sigma\backslash\tau} > 0$)

The last comment is of particular importance. The fact is that when the "*conventional failure of a regular mechanical object*" ($\omega_\Sigma = 1$) occurs, i.e., it disintegrates, at least, into two parts, the existence of the MTD system does not end—in accordance with Item XIV, a long period comes when an object disintegrates into particles ($1 < \omega_\Sigma^* \leq \infty$). Here, not so much mechanical loads, as electrochemical and thermodynamic phenomena (processes) are the determining parameters. On the basis of the above-said, the *law of any disintegration* (*decomposition*) *of the MTD system* is formulated in the form

$$\sum m_{V_{ijT}} = m_{V_0}. \tag{2.82}$$

Law (2.82) suggests the mass conservation of the system regardless of the conditions of its degradation and disintegration. In other words, the *mass of disintegrating particles* $\sum m_{V_{ijT}}$ (whatever their size) *cannot exceed* (or it can be less than) the *initial mass* m_{V_0} of *the MTD system*.

Hence, there is the need of the analysis (at least, short) of the *evolution of systems*.

2.9 Evolution of a System by Damageability

Give the description according to Tribo-Fatigue [39] of the behavior of the *deformable solid-solid system* in some environment on the most general—dialectical grounds. The *origination* of the system, *existence (occurrence, life* and *degradation)* can be represented in the following general form:

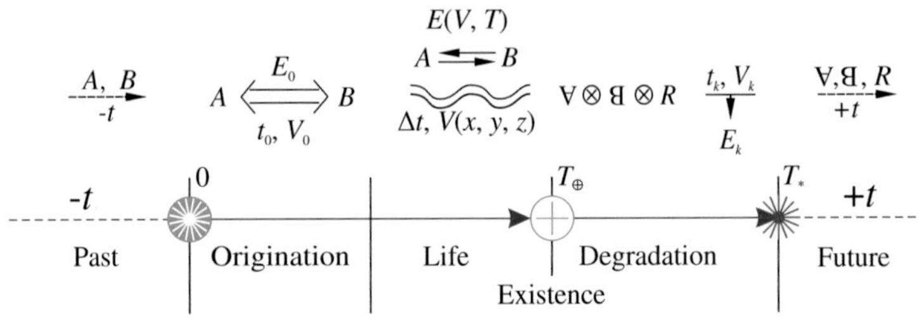

$$
\begin{array}{c}
E(V,T) \\
A \rightleftharpoons B
\end{array}
$$

$$
\underset{-t}{\overset{A,\ B}{\longrightarrow}} \qquad
A \underset{t_0,\ V_0}{\overset{E_0}{\Longleftrightarrow}} B \qquad
\underset{\Delta t,\ V(x,y,z)}{\overset{A \rightleftharpoons B}{\sim\!\sim}} \qquad
V \otimes \text{B} \otimes R \qquad
\underset{E_k}{\overset{t_k,\ V_k}{\downarrow}} \qquad
\underset{+t}{\overset{V,\text{B},\ R}{\longrightarrow}} \tag{2.83}
$$

Here *A, B* are some separate bodies (elements, etc.), details, objects. Their existence in the past $(-t)$ is sketched by the dashed line with an arrow.

This writing

$$
A \underset{t_0,V_0}{\overset{E_0}{\Leftrightarrow}} B \tag{2.84}
$$

means that the *creation (origination) of a system (or an object)* is the product of energy (E_0) interaction (\leftrightarrow) of bodies *A, B* implemented in the time t_0 within the volume V_0. Of course, this product is neither *A* nor *B*; but it is an entity with special integral characteristics and functions which neither *A* nor *B* can possess.

The writing

$$
\begin{array}{c}
E(V,T) \\
A \rightleftharpoons B \\
\underset{\Delta t,\ V(x,y,z)}{\sim\!\sim}
\end{array} \tag{2.85}
$$

means that the *life of the system* is the process of its energy $E(V,t)$ interaction with the environment $V(x,y,z)$ during the time Δt. This interaction with the environment always causes surface damages to originate and accumulate in the system elements since t, V, E are variable. The system itself is also characteristic for the force interaction of its elements $(A \rightleftharpoons B)$. This means is that not only surface damage, but also volumetric (internal) damage should arise and develop, since the forces of such interaction are distributed over the volume of the elements and vary with time. Therefore the life (longevity) of the system is shown in (2.83) and (2.85) by the wavy lines. *Accumulation of irreversible surface and volumetric damages is the softening process which eventually causes the system degradation and death.*

Assume that the *system elements*, as well as the *entire system reveal the hardening property*, i.e., the ability to increase its resistance by both external and internal influences when they are hardened. Then *the outcome* of *struggle of opposites* (i.e., *hardening-softening processes*) also *determines the life of the system, or its existence time (operation)*. If the damageability level grows in time, then the system degrades inevitably, as soon as the damageability reaches some limiting (or critical) value.

Thus the writing

$$Ɐ \otimes ꓭ \otimes R \quad \underset{E_k}{\overset{t_k,\, V_k}{\downarrow}} \tag{2.86}$$

means the following. The *degradation of the system* is the process that leads to its disintegration within the volume V_k in the time t_k into fragments (Ɐ, ꓭ) and residues R. The degradation is accompanied by the release of the energy E_k; the fragments and residues dispersed in time and space constitute a set (⊗) of disintegration products.

The *products of system degradation* are represented by three components in expressions (2.83) and (2.86). First, these are the modified bodies A and B (denoted as Ɐ and ꓭ, respectively). Secondly, these are system residues (denoted as R). In other words, Ɐ and ꓭ) are the recognizable parts (fragments) of the disintegration products of the system since A and B are their images. As far as R is bodies designated as Ɐ and ꓭ). Second, these are the residuals of the system designated as R. In other words, Ɐ and ꓭ) are recognizable parts of disintegration products since A and B are their images. As for R, it is the unrecognizable (or hardly recognizable) part of the disintegration products of the system. This part can be represented as the one consisting of at least four components:

$$R = R\,(R_A^{ꓭ}, R_B^{Ɐ}, R_{A,Ɐ}^{t,V}, R_{B,ꓭ}^{t,V}), \tag{2.87}$$

i.e., the $R_A^{ꓭ}$'s are the residuals A embedded in ꓭ and trapped by it. The $R_B^{Ɐ}$'s are the residuals B in Ɐ, i.e., these are the fragments B embedded in Ɐ and trapped by it. The $R_{AⱯ}^{t,V}$'s are the residuals A and Ɐ dissipated in the space (environment) V and in the time t. Finally, the $R_{B,ꓭ}^{t,V}$'s are the residuals B and ꓭ dissipated in the space V and in the time t.

Residuals and fragments are going in the future ($+t$). Their existence is shown in (2.83) by the dashed arrow. This existence can be separate and is marked by the commas between the symbols Ɐ, ꓭ, R.

Expression (2.83) should be understood as the conventional (symbolic) writing of the sequence of interrelated processes of system origination, existence, and degradation.

As the simplest specific example, consider one of the widespread active systems: *crankshaft journal (A)—sliding bearing (B) of the rod head of the engine*. Of interest is the life of the system.

The technological process of manufacturing parts A and B ends in the assembly $A \Leftrightarrow B$—it is the process of system origination (2.84). Obviously, it is implemented in the time t_0 within the volume V_0 at the energy expenditure E_0. Then system life (2.85) begins: aging, normal operation, gradual loss of efficiency. In the course of life the system $A \Leftrightarrow B$ changes into $A \leftrightarrows B$), i.e., assembly components undergo wearing at the contact pressure q and wear-fatigue damages accumulate in the crankshaft journal when acted upon by cyclic stresses σ. This occurs when the energy $E(V, t)$ interacts with the environment (oxidation of friction surfaces) during the entire existence time Δt. Thus, both the environment $V(x, y, z)$ and the interaction energy E vary with time. The damage accumulation causes the system to degrade according to (2.86) and, hence, its failure (wear-fatigue fracture of the crankshaft journal, frictional fracture of bearing inserts). The system undergoes failure in the environment V_k in the time t_k followed by the release of the energy E_k. In the process of failure (2.86), the fragments $\not{V}$ and $\not{B}$ are the parts of the shaft A and the inserts B. Also, the residuals R—wear products (2.87)—are formed: the crankshaft particles embedded into the sliding bearing inserts ($R_A^{\not{B}}$); the insert particles embedded into the crankshaft journal surface ($R_B^{\not{V}}$); the products of surface damage of the crankshaft journal ($R_{A\not{V}}^{t,V}$) and the inserts ($R_{B,\not{B}}^{t,V}$) scattered in the environment in the time t, i.e., the wear products removed from the friction zone.

As seen, based on (2.83), a sufficiently general and correct qualitative analysis of interactions of the system elements and the system with the environment is given.

The outlined qualitative picture can serve as a basis, for instance, for setting and describing quantitatively the life N (resource) of the active system. It is obvious that $\Delta t = N$ is the function of cyclic stresses σ in the crankshaft journal, the contact pressure q in the tribo-coupling, the wear rate I of system elements, the accumulation rate of wear-fatigue damage ϑ, the properties (composition, structure) of the environment C_V and the elements A, B of the system (C_A, C_B):

$$N = N(\sigma, q, I_\sigma, \vartheta, C_V, C_A, C_B, \ldots).$$

This equation for longevity can be specifically implemented, for instance, using the methods of applied mechanics.

Similarly, the processes of origination, life, and degradation of other systems, for example, *solid–fluid,* etc. can be described. Differences will be only in specifying what interaction forces are implemented in the investigated case and what damages arise and develop.

If *the biological system*, for example, cardiovascular or musculoskeletal is considered, then a sufficient qualitative description of its life, damage, and degradation can be made with the use of symbolic model (2.83) developed as applied to inorganic active systems. Further, it is necessary to take into account *a specific complex of biological phenomena and factors* [39]. It is shown that approach (2.83) can also be used to describe the general processes of birth, life, and death of *a living organism* that together with environment and habitat conditions forms the most complex living system in it. For this case, *the concept of Tribo-Fatigue life as a special method of damage accumulation* [36] is developed.

Table 2.5 Characteristics of damageability evolution of the MTD system

MTD system states		Parameters		State properties (physical)	State symbols	Energy conditions of states	Technogenic situations and possible damages
Symbol	Characteristic	Damageability	Integrity ($\delta = 1 - \psi$)				
1	2	3	4	5	6	7	8
A	Undamaged	$\omega_A = 0$	$\delta_A = 1$	Maintaining the integrity (size, shape, mass), structures (skeleton) and support (implementation) of all functions	$V_0 = \text{const}$ $\overset{A_0}{\underset{B_0}{\Leftrightarrow}}$ $u_\Sigma^{eff} = 0$	$u^{eff} = 0$ $\psi_u^{eff} = 0$	Failures (e.g., short-time reversible change of function)
B	Damaged	$0 < \omega_B < 1$	$1 > \delta_B > 0$	Development of complex damageability and malfunctioning	$V_{ij} > 0$ $\overset{A}{\underset{B}{\rightleftarrows}}$ $u_\Sigma^{eff} > 0$	$u_\Sigma^{eff} < u_0$ $\psi_u^{eff} < 1$	Incidents (e.g., permissible system wear)
C	Critical (limiting)	$\omega_\Sigma = 1 = \omega_C$	$\delta_C = 0$	Total functional loss, multicriterion limiting state	$C \in (V \otimes B)$	$u_\Sigma^{eff} = u_0$ $\psi_u^{eff} = 1 = \psi_C$ $d_c = 1$	Accidents (e.g., fatigue failure of engine shaft)
D	Supercritical	$1 < \omega_D^* < \infty$	$\delta_D < 0$	Formation of multiple fragments, dissipated fragments and residuals	R_A^B, R_B^V, $R_{AV}^{t,V}$, $R_{B,B}^{t,V}$	$1 > d_D^* > d$	Catastrophes (e.g., mid-air collision)
E	Disintegration (breakdown)	$\omega_E^* \to \infty$	$\delta_E \to -\infty$	Formation of nanoclusters, scattered atoms, elementary particles	✹	$d_E^* \to d_\hbar$	Cataclysms (e.g., nuclear explosion)

Approach (2.83) is also used for description of the evolution of the MTD system, including in the translimiting state. Table 2.5 contains this approach with regard to the above-described diverse characteristics of system damage. It is obvious that the qualitative representation (2.83) of the evolution is supplemented here with the specific numerical analysis—at all nodal points of development (states *A, B, C*) and degradation (states *C, D, E*).

The general classification of the conceivable states of a system (object) in terms of damage is contained in columns 1, 2, 3. It is similar to Table 2.4, but with the specification (as marked above) that the level of critical damageability (ω_Σ^*) is assigned the superscript that means such a state. Table 2.5 also contains the appropriate physical characteristics of system states (column 5) and the additional analysis (column 4) based on the characteristic of its integrity $(\delta = 1 - \omega_\Sigma)$. Column 6 contains the symbolic description of all system states. The above-described energy states of the system are based on conditions (2.7), (2.8) contain two uncertainties. These uncertainties are interpreted as follows. When $\psi_D^* \to \infty$ (according to condition (2.7), the absolute average size (d_ψ^*) of particles forming during the system decomposition must become arbitrarily small $(d_D^* \to 0)$ by condition (2.8). Table 2.5 reveals these uncertainties (column 7). Namely, it is assumed that transmitting states are described by the changes in the size of particles forming within the range

$$1 > d_D^* > (1/k),$$

where the left constrain is defined by unity (as a symbol of "*integrity*") and the right one—by the arbitrarily (or infinitely) *large integer k* such as within the limit

$$\lim_{k \to ⋔} (1/k) = \min d_D^* = d_{⋔} \approx 10^{-k}, \tag{2.88}$$

where the *conventional, yet finite* quantity *big* (⋔) is introduced as the limit of a possible growth of the integer *k* to the quantity (k = ⋔) that can be specified as the total quantity of atoms in the system under investigation. In principle, it can be calculated if the size $d_{⋔}$ of atoms is known for materials, of which the system is "made"; thus $d_{⋔} \approx 10^{-k}$. In (2.88) it is then considered that the *system death means its disintegration into such a "quantity" of particles that is equal to the initial number of atoms available in the system.* The latter can be reasonably calculated practically for any systems. It has been established, for example, that the amount of atoms in the Universe approximately equals 10^{67} [10].

Thus, the growth of the *level of translimiting damageability* of a solid $\omega_\Sigma^*, \psi_u^* > 1, d^* < 1$ (column 7, Table 2.5) signifies an appropriate decrease in the characteristic size of forming particles. Thus, the "location" of these particles is not specified—it can be any. But, naturally, it is meant that all particles will be finally spent for construction of those or other new systems (i.e., not necessarily—one system) [37]. This means that the *reproduction of systems* is inevitably

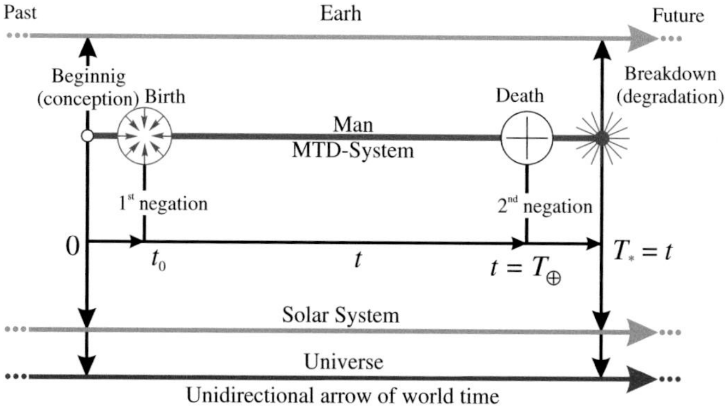

Fig. 2.23 Existence time of the material system

implemented *after their degradation*—but, of course, in new conditions with new initial parameters.

Note: a significant drawback of the performed analysis is the absence of the determining parameter—the time t.

As applied to the specific MTD system, Fig. 2.23 explains our idea of its existence. The general concept of the *unidirectional time arrow* is borrowed from thermodynamics (mostly from physics). Thus, the question about the *nature of time* is not being discussed here (as in physics and philosophy). Further, according to Item XIV, it is assumed that the *existence time of the system under examination is always finite* and is defined by the interval $(0, T_*)$ where T_* is the time before the system disintegration (Table 2.4). Within this interval the time *of its disintegration (failure)* $t = T_\oplus, T_\oplus \ll T_*$ is really defined. The failure of the system is interpreted as usual: it means the total loss of system functions and properties, which corresponds to the fact that the damageability measure (for example, ω_Σ or ψ_Σ) reaches the limiting (critical) value $\omega_c = 1 = \psi$. At the moment of failure the system, therefore, ceases to exist as a whole. Figure 2.23 shows that the existence of the system under study corresponds to a certain time interval on its any more general scale—for the Earth, the Solar system, the Universe (it is marked by the vertical arrows which separate the *past* and the *future*).

Now, describe the evolution of the MTD system.

Figure 2.24 illustrates that based on the mechanothermodynamic viewpoint, the A-evolution in time (Table 2.4) is implemented in *two stages*. The stage ABC $(\omega_\Sigma = \psi_u^{eff} \to 1)$ is the the *existence time of the system as* the *integrity* when it performs all its functions. It is represented as the *development accompanied by an inevitable growth of damage and deterioration of some functions* up to the moment when the limiting (critical) state is reached at the point S. *At this point the system completely "loses" all its functions*, for example, the accident (either the disintegration of one of the system elements into two parts, or the unacceptable (limiting) wear in system, etc.). *The second stage CDE then occurs and is represented as the*

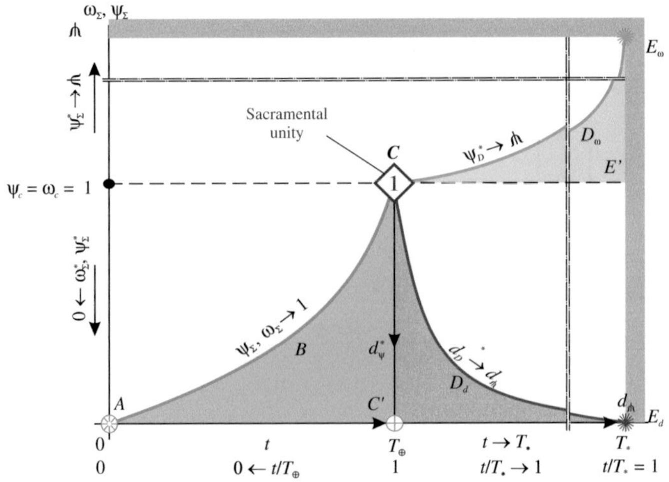

Fig. 2.24 MTD system evolution hypothesis

degradation process accompanied by the occurrence of numerous and various translimiting states caused, for example, by catastrophes, explosions, cataclysms, etc. Example: at the point C the vessel under static pressure is divided into 2 parts; it disintegrates into a large number of “infinitely small” particles, if the nuclear explosion (point E) is implemented in it; it collapses into fragments if the explosion in it is initiated by a different quantity of explosive substances (line CDE). The moment of disintegration of a solid into atoms (elementary particles, etc.) is denoted by the symbol (✸).

The second stage (*translimiting states*) can be described in two ways. Either the analysis of the *average size* $d_D^* \to d_{\pitchfork}$ of *disintegration particles* is used [formulas (2.80), (2.81)] and represented by the curve CD_dE_d in Fig. 2.24 [note that in this case, the origin of coordinates is shifted to the point C and the size d_D^* ranges from 1 to 0 (line CC')], or the damageability analysis $\psi_u^{eff} \to \pitchfork$ is used and represented by the curve $CD_\omega E_\omega$ in Fig. 2.24 (remind that here the “number of damages” corresponding to the disintegration (breakdown) of the system is designated by the number ψ conventionally equal to the number of atoms in the system.

The potentiality of the *parametric analysis* appears to be interesting and beneficial.

In our opinion, the representations as set forth above do not contradict the known and approved theories and the experimental results.

From Fig. 2.24 it is possible to find *two important features* of the time A-evolution of systems with respect to damageability.

First feature: the plot reveals the *sacramental point*, at which three special—critical units $\psi_u^{eff} = \omega_\Sigma = 1 = \omega_c, t/T_\oplus = 1$ and $d_\psi = 1 = d_c$ “come together”. It is the *evolution epicenter*, or its *apotheosis*. These critical units also define the “*division*” of A-evolution into two essentially differing stages—*development stage*

ABC and *degradation stage CDE*. It is the point when the system loses all its functions, i.e., the point of transition to various translimiting states.

Second feature: using the plot in Fig. 2.24, it is obviously possible to describe and define the *effective energy conservation law*.

$$\int_{0}^{T_{\oplus}} u_{\Sigma}^{eff}(t)dt \equiv \int_{T_{\oplus}}^{T_{*}} u_{eff}^{\Sigma}(t)dt, \tag{2.89}$$

where $u_{\Sigma}^{eff}, u_{eff}^{\Sigma}$ is the effective energy on the first (development) and on the second (degradation) stages, respectively. The statement of this law is as follows: *effective energy absorbed by the system in the process of reaching the limiting (critical) state is identically equal to the released (scattered) effective energy in the process of degradation up to disintegration* (for example, into atoms).

Geometrically, this law requires the equality of the three areas in Fig. 2.24

$$S_{ABCC'} \equiv \varphi S_{C'CD_dE_d} \equiv S_{CD_{\omega}E_{\omega}E'}, \tag{2.89a}$$

where ϕ is the function of parameter transformation [for example, according to (2.81)].

Identities (2.89), (2.89a) express the *so-called sufficiency principle*. If it is violated, then it means that the energy u_{Σ}^{eff} supplied to the system prior to its limiting state (u_0) was larger than it was needed to achieve the mentioned state, i.e., it was *excessive* $(u_{\Sigma}^{eff} > u_0)$.

From the above-stated, three main conclusions follow:

1. *Damages are the fundamental physical property (and the functional duty) of any system and all of its elements.*
2. *Damageability of each object (any existing one) inevitably grows up to its breakdown—decomposition (disintegration) into a set of particles of arbitrarily small size, i.e., it is the unidirectional process of time:*

$$\left\{\begin{array}{l} \Psi_{\Sigma}^{U} = \Psi_{\Sigma}^{U}\left(\sigma_n^{(V,W)}, \varepsilon_n^{(V,W)}, T_{\Sigma}, V_{ij}, V_{ijT}, V_0, Ch, \Lambda_{i\backslash j}, m_k, t, u_0\right) \Rightarrow \pitchfork, \\ \\ \end{array}\right. \tag{2.90}$$

$$\vec{d}_{D} \overset{\vec{t}}{\Rightarrow} d_{\pitchfork}. \tag{2.91}$$

3. *Not only the unity and struggle of opposites but also the directivity of various and complex physical processes of hardening-softening (depending on the level of loads and time) are typical of the system evolution by damageability. It means that the Λ-function of damage interactions (of all kinds) can take three classes of values:* (1) $\Lambda < 1$ when the hardening process is dominant; (2) $\Lambda > 1$ when the

softening process is dominant; (3) $\Lambda = 1$ when a stable hardening-to-softening process ratio is found.

References

1. Bogdanovich, A.V.: Prediction of Limiting State of Active Systems. Press of GrSU named after Ya.Kupala, Grodno (2008). (in Russian)
2. Cherepanov, G.P.: Fracture mechanics and kinetic theory of strength. Strength Mater. **11**, 3–8 (1989). (in Russian)
3. Chizhik, S.A., et al.: Specific features of generation and development of small surface cracks in carbon steel at contact-mechanical fatigue. Factory Lab. Diagn. Mater. **3**, 34–38 (1996)
4. Fridel, J.: Dislocations. Mir, Moscow (1967) (Russian Edition)
5. Gabar, I.G.: Analysis of the failure of BCC- and FCC-metals and the concept of interrelation of fatigue curve parameters. Strength Mater. **11**, 61–64 (1989). (in Russian)
6. Giurgiutiu, V., Reifsnider, K.L.: Development of strength theories for random fiber composites. J. Compos. Tech. Res. **16**(2), 103–114 (1994)
7. Hibbeler, R.C.: Technical Mechanics 2—Strength Theory 5. Munich (2005)
8. Ivanova, V.S., Terentiev, V.F.: Nature of Metals Fatigue. Metallurgiya, Moscow (1975). (in Russian)
9. Kittel, C.: An Introduction to Solid Physics. Nauka, Moscow (1978). (in Russian)
10. Kotlin, A.: Any Infinity is Finite. http://www.proza.ru/2010/04/03/1157
11. Yu, M.-H.: Unified Strength Theory and Its Applications. Springer, Berlin (2004)
12. Oleynik, N.V.: Fatigue Strength of Materials and Machines Components in Corrosive Media. Naukova dumka, Kiev (1987). (in Russian)
13. Pisarenko, G.S., Lebedev, A.A.: Deformation and Strength of Materials in Complex Stress States. Naukova dumka, Kiev (1976). (in Russian)
14. Ponomarev, S.D., et al.: Strength Calculations in Mechanical Engineering. State Scientific and Technical Publishing House of Machine-Building Literature, Moscow (1958). (in Russian)
15. Prokopchuk, N.V.: Temperature dependence of activation energy of mechanical failure of polymeric materials. Strength Mater. **10**, 46–50 (1984). (in Russian)
16. Regel, V.R., Slutsker, A.I., Tomashevskii, E.E.: Kinetic Nature of the Strength of Solids. Nauka, Moscow (1974). (in Russian)
17. Serensen, S.V.: Problem of fatigue and wear resistance of details of machine components (brief overview). Increase Wear Resist. Lifetime Mach. **1**, 10–14 (1960). (in Russian)
18. Sherbakov, S.S., Sosnovskiy, L.A.: Mechanics of Tribo-Fatigue Systems. BSU Press, Minsk (2010). (in Russian)
19. Sherbakov, S.S., Sosnovskyi, L.A., Zhuravkov, M.A.: Statistical model of a deformable solid. In: Strength of Materials and Construction Elements, International Scientific Technical Conference, Kiev, vol. 2, pp. 207–209. Press of the G.S. Pisarenko IPP NAS Ukraine, Kiev, 28–30 Sept 2010
20. Shultze, G.: Metallophysics. Mir, Moscow (1971) (Russian Edition)
21. Sitamov, S., Kartashov, É.M., Khukmatov, A.I.: Processes of failure of polymers in various types of loading in the brittle and quasibrittle state. Strength Mater. Factory Lab. Diagn. Mater. **1**, 37–40 (1989). (in Russian)
22. Sosnovskiy, L.A., et al.: Reliability, Risk, Quality. BelSUT Press, Gomel (2012). (in Russian)
23. Sosnovskiy, L.A. (ed.): Tribo-Fatigue–98/99. Issue 1: Theory of Accumulation of Wear-Fatigue Damage. Press of SPA "TRIBO-FATIGUE", Gomel (2000) (in Russian)
24. Sosnovskiy, L.A., Gapanovich, V.A.: Principal ways for improving operational reliability of Tribo-Fatigue system brake block-wheel-rail. In: Zhuravkov, M.A., et al. (eds.) Proceedings of

6th International Symposium on Tribo-Fatigue, Minsk. Part 2, pp. 187–201. BSU Press, Minsk, 25 Oct–1 Nov 2010 (in Russian)
25. Sosnovskiy, L.A., Sherbakov, S.S.: Mechanothermodynamic system and its behavior. Continuum Mech. Thermodyn. **24**(3), 239–256 (2012)
26. Sosnovskiy, L.A., Sherbakov, S.S.: Possibility to construct mechanothermodynamics. Nauka i Innovatsii **60**, 24–29 (2008). (in Russian)
27. Sosnovskiy, L.A., Sherbakov, S.S.: Surprises of Tribo-Fatigue. Magic Book, Minsk (2009)
28. Sosnovskiy, L.A.: Fundamentals of Tribo-Fatigue. BelSUT Press, Gomel (2003). (in Russian)
29. Sosnovskiy, L.A.: Mechanics of the irreversible damages caused by contact and noncontact load. In: Proceedings of World Tribology Congress III, pp. 2–8, Washington, 12–16 Sept 2005
30. Sosnovskiy, L.A.: Tribo-Fatigue. Wear-Fatigue Damage and Its Prediction (Foundations of Engineering Mechanics). Springer, Berlin (2005)
31. Sosnovskiy, L.A.: Fundamentals of Tribo-Fatigue. BelSUT Press, Gomel (2003). (in Russian)
32. 索斯洛夫斯基著, L.A.: 摩擦疲劳学 磨损—疲劳损伤及其预测. 高万振译—中国矿业大学出版社 (2013) (in Chinese)
33. Sosnovskiy, L.A.: L-Risk (Mechanothermodynamics of Irreversible Damages). BelSUT Press, Gomel (2004). (in Russian)
34. Sosnovskiy, L.A., Makhutov, N.A.: Methodological problems of a comprehensive assessment of damageability and limiting state of active systems. Factory Lab. **5**, 27–40 (1991)
35. Sosnovskiy, L.A., Sherbakov, S.S.: Surpises of Tribo-Fatigue. BelSUT Press, Gomel (2005). (in Russian)
36. Sosnovskiy, L.A.: Life field and golden proportions. Nauka i Innovatsii **79**(9), 26–33 (2009) (in Russian)
37. Sosnovskiy, L.A.: Mechanics of Wear-Fatigue Damage. BelSUT Press, Gomel (2007). (in Russian)
38. Sosnovskiy, L.A.: Statistical Mechanics of Fatigue Damage. Nauka i Tekhnika, Minsk (1987). (in Russian)
39. Sosnovskiy, L.A.: Tribo-Fatigue: About Dialectics of Life. S&P Group TRIBOFATIGUE, Gomel (1999)
40. Sysoev, P.V., Bogdanovich, P.N., Lizarev, A.D.: Deformation and wear of polymers at friction. Nauka i Tekhnika, Minsk (1985)
41. Timoshenko, S.P., Goodier, J.: Theory of Elasticity. Nauka, Moscow (1975)
42. Troschenko, V.T., et al. (eds.): Strength of Materials and Constructions. Akademperiodika, Kiev (2005) (in Russian)
43. Troshchenko, V.T.: Fatigue and Inelasticity of Metals. Naukova dumka, Kiev (1971)
44. Vasilevich, Y.V., Podskrebko, M.D., Neumerzhizkaya, E.Y.: Statistical methods in calculations of construction elements for strength and reliability. In: Zhuravkov, M.A., et al. (eds.) Proceedings of 6th International Symposium on Tribo-Fatigue, Minsk. Part 1, pp. 229–232. BSU Press, Minsk, 25 Oct–1 Nov 2010
45. Vysotskiy, M.S., Vityaz, P.A., Sosnovskiy, L.A.: Mechanothermodynamic system as the new subject of research. Mech. Mach. Mech. Mater. **15**, 5–10 (2011). (in Russian)
46. Zhuravkov, M.A., Sherbakov, S.S.: Dangerous volumes in the active system. In: 10th Belarusian Mathematical Conference, Minsk, BSU. Part 2, pp. 121–122. Press of the Institute of Mathematics of NAS Belarus, Minsk, 3–7 Nov 2008
47. Zhuravkov, M.A., Sherbakov, S.S.: Analysis of damageability of an active system using the model of a deformable solid with a dangerous volume. Theor. Appl. Mech. **25**, 44–49 (2010)
48. Zhurkov, S.N.: Dilaton mechanism of the strength of solids. Phys. Strength Solids **1**, 5–11 (1986). (in Russian)
49. Zhurkov, S.N.: Kinetic concept of the strength of solids. Proc. USSR Acad. Sci. **3**, 46–52 (1968). (in Russian)

Chapter 3
Entropy States of Mechanothermodynamic Systems and Their Evolution

Abstract The theory of evolution is stated on the basis of the entropy concept. In addition to the traditional understanding of (thermodynamic) entropy as a characteristic of energy dissipation, the concept of Tribo-Fatigue entropy as a characteristic of energy absorption is introduced. The analysis of time changes in total (joint) entropy leads to the law of increase in entropy of any systems.

3.1 General Notions

The entropy approach to the analysis of the states of *mechanothermodynamic* systems is given below. The sequence of our analysis is as follows. First, we introduce the *concept of Tribo-Fatigue entropy* generated in a mechanical system similar to *thermodynamic entropy* determined by energy and substance exchange. The radical difference of these concepts is the following: *thermodynamic entropy* is a characteristic of energy dissipation, while, *Tribo-Fatigue entropy* is a *characteristic of its absorption, hence, damage of moving and deformable solids*. Combining these concepts allows general outlines of *mechanothermodynamics* in terms of entropy to be constructed. To understand *the evolution of a system,* one needs to establish an interrerelation of *motion, damage, and information*. It is shown that *the motion generates new information* in a system *if its damageability index is nonzero*; *the information appears to be positive when the system is hardened and negative when it is softened*. This gives an impetus to perceive *the fundamental peculiarity of interaction of irreversible damages* (effective energy, entropy) of different-nature (mechanical loads, thermal flows, etc.): the interaction reveals the *dialectical character* (so-called Λ-*interactions*). It has appeared that the *damageability is* the *fundamental physical property* (*duty*) *of a mechanothermodynamic system* and Λ-interactions determine its evolution from the viewpoint of damageability having regard to diverse and complex hardening-softening processes ($\Lambda \gtreqless 1$). It has appeared that Λ-interactions define the evolution of entropy states of a system, whose main regularity reads: *entropy of MTD systems inevitably grows in time.*

L. Sosnovskiy and S. Sherbakov, *Mechanothermodynamics*,
DOI 10.1007/978-3-319-24981-0_3

3.2 Tribo-Fatigue Entropy

To describe the state of thermodynamic systems, the following functions of internal energy U and entropy S

$$U = U(T, V, N_k) \text{ or } S = S(T, V, N_k), \tag{3.1}$$

are used where temperature T, volume V, and number of mols of chemical components N_k are the macroscopic state variables.

In the general case of an open system, the change dU of internal energy U is presented [5] as

$$dU = dQ + dA + dU_{\text{sub}} = TdS - pdV + \sum_{1}^{n} \mu_k dN_k, \tag{3.2}$$

where dQ is the amount of heat; dA is the amount of mechanical energy; dU_{sub} is the amount of substance, with which the system exchanged with the environment for a time interval dt; p is the pressure; the μ_k's are the chemical potentials; $dN_k = d_iN_k - d_eN_k$ is the change in the number of mols due to irreversible chemical reactions and substance exchange with the environment. *Max Planck* especially emphasized that in formula (3.2) dU is an infinitely small difference, whereas, dQ, dA, and dU_{sub} are the infinitely small amounts.

From (3.2) it follows that the *entropy change* is

$$dS = \frac{dU + pdV}{T} - \frac{1}{T}\sum_{1}^{n} \mu_k dN_k. \tag{3.3}$$

Entropy increment (3.3) according to *Prigogine* can be presented as the sum of its change $d_eS \gtreqless 0$ due to *system and energy exchange and due to substance and environment exchange* and the change $d_iS \geq 0$ due to irreversible processes occurring inside the system:

$$dS = d_eS + d_iS. \tag{3.4}$$

Thus, in *thermodynamics, the entropy S is a measure of irreversible energy dissipation* [11] that characterizes the system state from the viewpoint of its internal order or structure.

Equations (3.2) and (3.3) do not take into account many processes, for example, internal energy changes at damage of moving and deformable solids and Tribo-Fatigue systems [16, 18, 19]. Substance exchange is considered only as a result of such processes as diffusion and chemical reactions, whereas substance exchange at surface wear and volume (e.g., wear-fatigue) damage is not allowed for. That is why, the problem on evaluating the entropy arises in relation to numerous phenomena of damageability. Such phenomena are characteristic of the systems with *moving and deformable objects*.

According to the generalized concepts [16, 18–20] given in Chap. 2, the damage is the change in composition, construction, structure, size, shape, volume, continuity, mass and, hence, in the corresponding physical–chemical, mechanical, and other properties of an object; finally, the damage is related to the discontinuity and integrity of a solid up to its decomposition (e.g., into atoms) [20]. Thus, the damageability is treated as a fundamental property (and duty) of moving and deformable systems [16, 18–20], and the failure is considered as a specific type of damage, i.e., the corresponding discontinuity and integrity of objects.

In Tribo-Fatigue, it is shown [16, 18, 19] that for Tribo-Fatigue systems, the *irreversible damageability* ω_Σ (as well as Ψ_u^{eff}) is the function of the effective mechanical U_M^{eff}, thermal U_T^{eff} and electrochemical U_{ch}^{eff} energies. Here, the mechanical energy due to changes in solid size (U_σ^{eff}) and due to changes in solid shape (U_τ^{eff}) is distinguished:

$$\begin{aligned}\omega_\Sigma &= \omega_\Sigma\left(U_\sigma^{eff}, U_\tau^{eff}, U_T^{eff}, U_{ch}^{eff}\right)\\ &= \omega_\Sigma\left(\sigma_{ij}, \varepsilon_{ij}, T_\Sigma, v_{ch}(m_v), \Lambda_{\sigma\backslash p}, \Lambda_{T\backslash M}\right) = \omega_\Sigma\left(U_\Sigma^{eff}\right).\end{aligned} \tag{3.5}$$

Here, the Λ-functions characterize the interaction of damages due to different loads (force and contact-friction denoted by the subscript σ/p; thermal and mechanical—by the subscript T/M), T_Σ is the temperature caused by all heat sources, $U_\Sigma^{eff} = U_\Sigma^{eff}\left(\sigma_{ij}, \varepsilon_{ij}, T_\Sigma, v_{ch}(m_v), \Lambda_{\sigma\backslash p}, \Lambda_{T\backslash M}\right)$ is the total effective energy.

In (3.5), the known interrelation of energy and corresponding force factors are taken (σ_{ij} and ε_{ij} are the stress and strain tensors, v_{ch} is the rate of electrochemical processes with regard to the material properties (m_v)). The energy directly spent for the formation and development of irreversible damages is called effective energy, i.e., U^{eff} is the absorbed part of the energy supplied to the system [16, 18, 19]. The methods of its determination are outlined in [16, 18–20] and provide the formulas for estimation of ω_Σ under different operating conditions of Tribo-Fatigue systems.

In Sect. 3.1, it is shown that irreversible damages are as a rule formed and accumulated not within the entire (geometric) volume of a deformable solid but only within its finite region with a critical state; this region is called the *dangerous volume*.

As internal irreversible damages of thermomechanical nature originate due to effective energy changes in the dangerous volume $V_{P\gamma}$ of the system, in the general case, we have

$$dU_\Sigma^{eff} = \gamma_1^{(w)} \omega_\Sigma dV_{P\gamma}, \tag{3.6}$$

where $\gamma_1^{(w)}$ is the stress (pressure) that causes the damage of a unit dangerous volume ($V_{P\gamma} = 1$).

Then, according to (3.2) and (3.4), it is possible to introduce the concept of *Tribo-Fatigue entropy*, whose change is

$$(d_iS)_{\mathrm{TF}} = \frac{\gamma_1^{(w)}}{T_\Sigma}\omega_\Sigma dV_{P\gamma}. \tag{3.7}$$

Thus, the *Tribo-Fatigue entropy serves as a measure of irreversible absorption of the energy U_Σ^{eff} in the dangerous volume $V_{P\gamma}$ of the Tribo-Fatigue system.*

Show the analogy between the concepts of Tribo-Fatigue and thermodynamic entropy.

According to thermodynamics, in the general case, the irreversible change of the entropy d_iS is associated with the flux of some quantity X (for example, heat or substance):

$$d_iS = FdX, \tag{3.8}$$

where F is the thermodynamic force. So, at gas expansion in the piston engine the pressure of the gas volume (p_1) is always larger than the piston pressure (p_2). The difference ($p_1 - p_2$) then characterizes the pressure gradient and is the force per unit area that displaces the piston. At $T =$ const, the irreversible entropy increment [5] is

$$(d_iS)_T = \frac{p_1 - p_2}{T} dV > 0. \tag{3.9}$$

Here, $(p_1 - p_2)/T$ corresponds to the thermodynamic force (F) and dV characterizes the thermodynamic flux (dX) associated with it.

If ω_1 is the damage concentration at the point of a solid with the largest stress (p_1) and ω_2 is the damage concentration at any solid point, at which the stress (p_2) is smaller, then $\omega_2 < \omega_1$, i.e., there is a damage gradient characterized by the difference ($\omega_1 - \omega_2$) that is associated with the value of the dangerous volume $V_{P\gamma}$.

The irreversible Tribo-Fatigue entropy increment at T_Σ = const is

$$(d_iS)_{TF} = \gamma_1^{(w)} \frac{\omega_1 - \omega_2}{T_\Sigma} dV_{P\gamma} > 0. \tag{3.10}$$

Thus, here $\gamma_1^{(w)}(\omega_1 - \omega_2)/T_\Sigma$ corresponds to the thermodynamic force F since $\omega_1 \sim p_1$ and $\omega_2 \sim p_2$, whereas $dV_{P\gamma}$ characterizes the thermodynamic flux X associated with this force.

Now consider an open thermodynamic system with a damageable solid; it is a mechanothermodynamic system. The entropy increment in such a system is obviously determined by the sum of thermodynamic entropy (3.3) and Tribo-Fatigue entropy (3.7):

$$(dS)_T + (d_iS)_{TF} = \left(\frac{dU + \Delta p dV}{T} - \frac{1}{T}\sum_1^n \mu_k dN_k\right)_T + \frac{\gamma_1^{(w)}}{T_\Sigma}\omega_\Sigma dV_{P\gamma}, \quad (3.11)$$

if to a second approximation, the both entropy components are assumed to be additive.

Here, thermodynamic entropy (3.3) has a subscript T; in this case, it is taken into account that $\Delta pdV = (p_M - p)dV$, $p_M dV$ is the mechanical energy supplied to the system from the environment. If $\omega_\Sigma = 0$, then (3.11) reduces to (3.2). Tribo-Fatigue entropy (3.7) is denoted by the subscript TF.

Equation (3.11) for the mechanothermodynamic state is *radically different* from Eq. (3.3) for the thermodynamic state: the first permits any state of the system to be analyzed, including *A*-, *B*-, *C*-, *D*- and *E*-states of damage (Table 2.4) since in the general case, $0 \le \omega_\Sigma \le \infty$ [16, 18–20]. Hence, according to (3.11), exactly the growth of Tribo-Fatigue entropy production (3.7) due to the thermomechanical state of the system can cause both its damage and decomposition; thermodynamic Eq. (3.3) does not take into account such states. But it must be admitted that the state of any system to a large extent depends, for example, whether some element present in the system will be destroyed or not. The element must be destroyed if the entropy production in it will become critical (for reasons, not discussed here). The problem of critical and supercritical entropy levels caused by damage and failure of the system is not yet studied [5].

Thus, *irreversible damage processes in the dangerous volume of the active system generate the Tribo-Fatigue entropy, the change of which ceases only after the system death.*

Generalizing the aforesaid, write (3.7) in terms of the matrix determinant L_{ω_Σ} of the system damageability:

$$(d_iS)_{TF} = \gamma_1^{(w)} \frac{L_{\omega_\Sigma}}{T_\Sigma} dV_{p\gamma}; \quad (3.12)$$

$$L\omega_\Sigma = \begin{vmatrix} \omega_{11} & \omega_{12} & \dots & \omega_{1n} \\ \omega_{21} & \omega_{22} & \dots & \omega_{2n} \\ \dots & \dots & \dots & \dots \\ \omega_{n1} & \omega_{n2} & \dots & \omega_{nn} \end{vmatrix}, \quad (3.13)$$

It can be composed of such a number of the components (ω_{ij}) that corresponds to the number of stress tensors that affect the changes in the damageability state of the system The concept of the *damage tensor* of type (3.13) was introduced in [16, 18–20]. In the general case, according to (3.5), we have

$$\omega_{ij} = \omega_{ij}\left(\sigma_{ij}, \sigma_*\right), \quad (3.14)$$

where σ_* is the characteristic limiting stress [24, 26].

Table 3.1 Thermodynamic flows and forces in some frequently observed irreversible processes

Phenomenon	Flow	Force	Quantity
Heat transfer	Heat flux J_{th}	$\nabla(1/T)$	Vector
Diffusion	Mass flow of component *i*, $J_{d,i}$	$-[\nabla(\mu_i/T) - F_i]$	Vector
Viscous flow	Dissipative part of pressure tensor *P*	$\nabla v(1/T)$	2nd rank tensor
Chemical reaction	Reaction rate ρ, ω_ρ	Reaction affinity divided by *T*, A_ρ/T	
Damage	Damage flow $J_{V_{P_\gamma}}$	$L_{\omega_\Sigma}(1/T_\Sigma)$	2nd rank tensor

Note *T* is the temperature; μ_i is the chemical potential of the component *i*; F_i is the outside force per unit mass of the component *i*; *v* is the hydrodynamic velocity
The affinity A_ρ is connected with μ_i by the relation
$A_\rho = -\sum_i \nu_{i\rho}\mu_i$,
where the stoichiometric coefficients ν_i yield a complete set of molecules formining ($\nu > 0$) or vanishing ($\nu < 0$) in the reaction

Now, the data [10] on thermodynamic flows and forces in some frequently observed processes can be supplemented (Table 3.1) with the Tribo-Fatigue entropy concept.

Thus, from the aforesaid, it follows that *in the general case, the life (or fate) of the system is defined by the rate of irreversible changes in entropy—thermodynamic and Tribo-Fatigue; the production of internal mechanothermodynamic entropy is as eternal, as motion, and damage.*

It should be noted that in *continuum mechanics* [6, 12] the *stress tensor is expanded into two components*:

$$\sigma_{ij} = \sigma_{ij}^{(c)} + \sigma_{ij}^{(d)},$$

where, the superscript (*c*) denotes the tensor of conservative stresses and the superscript (*d*)—the tensor of dissipative stresses.

Then, the appropriate energy analysis yields the following *thermomechanical function of entropy*

$$\frac{dS}{dt} = \frac{1}{T}\frac{dq}{dt} + \frac{1}{\rho T}\sigma_{ij}^{(d)}\dot{\varepsilon}_{ij}, \tag{3.15}$$

where *dq*/*dt* is the rate of the heat flux to the medium per unit mass; $\frac{1}{\rho}\sigma_{ij}^{(d)}\dot{\varepsilon}_{ij}$ is the energy dissipation rate per unit mass (ρ is the continuum density).

Equation (3.15) is valid only for a continuum, provided that the procedure of dividing the stress tensor by dissipative and conservative parts is known. If the continuity of a deformable solid is disturbed, it cannot be used, i.e., it is unable to describe critical and supercritical states of the system, for example, according to Table 2.4. This is just the cardinal difference between Eqs. (3.15) and (3.11).

Moreover, in (3.15), we are talking about the entropy caused by static deformation, whereas in proposed model (3.7)—about the entropy generated by the processes of wear-fatigue damage under the action of an arbitrary system of loads (volumetric cyclic deformation, friction, etc.). At last, if (3.7) is considered in generalized form (3.12), then the Tribo-Fatigue entropy is the 2nd rank tensor, whereas thermomechanical entropy (3.15) is the scalar.

It should be noted that *function* (3.15) *has no community necessary for MTD systems*. For example, it does not allow the entropy to be estimated at its numerous its sources. So, if the friction and wear processes are realized in the system, then function (3.15) is weak. Or, if fatigue (mechanical, thermomechanical, etc.) processes are realized in the system, then function (3.15) again appears to be weak. Then we have to look for special (additional) characteristics of the entropy generated by fatigue or wear [1–3, 7–9]. Vice versa, *formula* (3.12) *possesses practically infinite community: it can be used for analyzing any phenomena and processes accompanied by the generation and development of irreversible damages of any nature.*

Thus, thermodynamics studies energy dissipation in systems; the (thermodynamic) entropy is a convenient characteristic of this process. If it is taken into account that internal irreversible damage is a fundamental property and a duty of a moving and deformable system, then it is not difficult to arrive at the concept of Tribo-Fatigue entropy (3.7), (3.12) as a measure of absorption of energy spent for generating and developing such damages in the mechanothermodynamic system. Generally, the latter is termed as an open (thermodynamic) system containing a moving and (cyclically) deformable solid. A specific feature of the mechanothermodynamic system, unlike the thermodynamic system, is that both thermodynamic and Tribo-Fatigue entropies are generated in it.

Unified model (3.11), as well as particular models (3.7), (3.12) predict the death of the system (for example, by its decomposition), if it undergoes damageability state evolution (3.5) described in Table 2.4; the analysis of such evolution is given in [20]. Particular model (3.3) is only the measure of disintegrating (or ordering) dissipative structures.

3.3 Calculation of Entropy

Consider, the example of entropy calculations for the mechanothermodynamic system containing a friction pair. Contact interaction is realized over the elliptic contact area, for which the smaller b-to-bigger a semi-axes ratio is $b/a = 0.574$. One of the elements of the friction pair is loaded by noncontact bending. An example of such an element is the shaft in the roller/shaft Tribo-Fatigue system.

Specific damageability Ψ of an elementary volume dV according to [13, 15, 16, 24] and by analogy with (2.5) can be presented as a relation between the parameter

ϕ_{ij} of the current mechanical state (stresses and strains) of the system and its limiting value $\phi_{ij}^{(\lim)}$.

Such relations may be of two kinds: dimensional and dimensionless for the stress tensor

$$\psi_{ij^*} = \sigma_{ij} - \sigma_{ij}^{(\lim)}; \tag{3.16}$$

and

$$\psi_{ij} = \sigma_{ij}/\sigma_{ij}^{(\lim)}, \tag{3.17}$$

for the stress intensity

$$\psi_{\text{int}^*} = \sigma_{\text{int}} - \sigma_{\text{int}}^{(\lim)}; \tag{3.18}$$

and

$$\psi_{\text{int}} = \sigma_{\text{int}}/\sigma_{\text{int}}^{(\lim)}; \tag{3.19}$$

where

$$\sigma_{\text{int}} = \frac{\sqrt{2}}{2}\sqrt{(\sigma_{11}-\sigma_{22})^2 + (\sigma_{22}-\sigma_{33})^2 + (\sigma_{33}-\sigma_{11})^2 + 6\left(\sigma_{12}^2+\sigma_{23}^2+\sigma_{13}^2\right)},$$

and the energy is

$$\psi_{u^*} = u - u^{(\lim)}; \tag{3.20}$$

$$\psi_u = u/u^{(\lim)}, \tag{3.21}$$

where the strain energy is

$$u = \sum_{i,j=1}^{3} \int_{\varepsilon_{ij}=0}^{\varepsilon^{(\lim)}} \sigma_{ij} d\varepsilon_{ij}, \tag{3.22}$$

and becomes

$$u = \frac{1}{2}\sum_{i,j=1}^{3} \sigma_{ij}\varepsilon_{ij}, \tag{3.23}$$

in the case of the considered linear dependence between stresses σ_{ij} and strains ε_{ij}.

According to [13, 15, 16, 18, 19, 24], the dangerous volume in a solid is a three-dimensional set of elementary volumes dV where acting stresses (strains), stress intensity, or strain energy surpass their limiting values and therefore produce the damageability of a solid:

$$
\begin{aligned}
V_{ij} &= \left\{ dV/\phi_{ij} \geq \phi_{ij}^{(\mathrm{lim})}, dV \subset V_k \right\}, \quad i,j = x,y,z; \\
V_{\mathrm{int}} &= \left\{ dV/\phi_{\mathrm{int}} \geq \phi_{\mathrm{int}}^{(\mathrm{lim})}, dV \subset V_k \right\}; \\
V_u &= \left\{ dV/u \geq u^{(\mathrm{lim})}, dV \subset V_k \right\},
\end{aligned}
\tag{3.24}
$$

where V_k. is the working volume.

Consider, the following expression for calculation of the energy dangerous volume V_u and the Tribo-Fatigue entropy S_u

$$
\begin{aligned}
V_u &= \iiint\limits_{u \geq u^{(\mathrm{lim})}} dV; \\
S_u &= \iiint\limits_{u \geq u^{(\mathrm{lim})}} dS_u dV,
\end{aligned}
\tag{3.25}
$$

where by the entropy definition, for example, according to [5] and with regard to expressions (3.3), (3.10), (3.20), the expression for the specific entropy per unit volume will be (accurate to constant)

$$
s_u = \frac{\psi_{u^*}}{T} = \frac{u - u^{(\mathrm{lim})}}{T}. \tag{3.26}
$$

If the damageability presented by the ratio $\psi_u = u/u^{(\mathrm{lim})}$, is used for calculating the Tribo-Fatigue entropy, then, the coefficient of appropriate dimension should be introduced.

Note that integrals (3.25) are calculated only in the dangerous volume where $u \geq u^{(\mathrm{lim})}$ and the energy is therefore absorbed to produce the damage unlike the volumes where the energy is just dissipated if $u < u^{(\mathrm{lim})}$. Calculation of these integrals due to the complexity of the surface bounding the dangerous volume was performed numerically using the Monte-Carlo method.

In the case of the contact interaction over the elliptical area, the pressure is expressed as

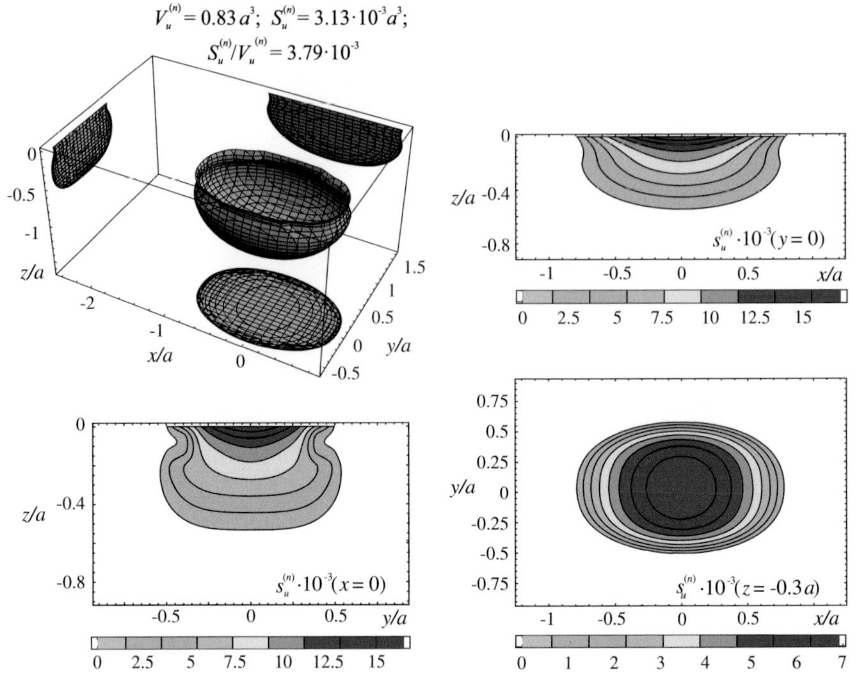

Fig. 3.1 Energy dangerous volume and its sections with specific entropy distributions for contact interaction without friction

$$p^{(n)}(x, y) = p_0^{(c)} \sqrt{(1 - x^2/a^2 - x^2/b^2)},$$

where $p_0^{(c)}$ is the maximum contact stress under the action of force F_c.

The entropy calculation in the Tribo-Fatigue system was based on the following initial data:

$$\begin{aligned} p_0^{(c)} &= \sigma_{zz}^{(n)}(F_c)\big|_{x=0,y=0,z=0} = 2960\,\text{MPa}, \\ p_0^{(c,\text{lim})} &= p_0\left(F_c^{(\text{lim})}\right) = 888\ \text{MPa} = 0.3p_0^{(c)} \end{aligned} \tag{3.27}$$

where $p_0^{(c,\text{lim})}$ is the contact fatigue limit (maximum contact stress under the action of the limiting force $F_c^{(\text{lim})}$ obtained in the course of mechano-rolling fatigue tests described in [13, 16, 18, 19, 24]. The criterion of the limiting state in these tests was the limiting approach of the axes in the Tribo-Fatigue system (100 μm). The test base was equal to $3 \cdot 10^7$ cycles.

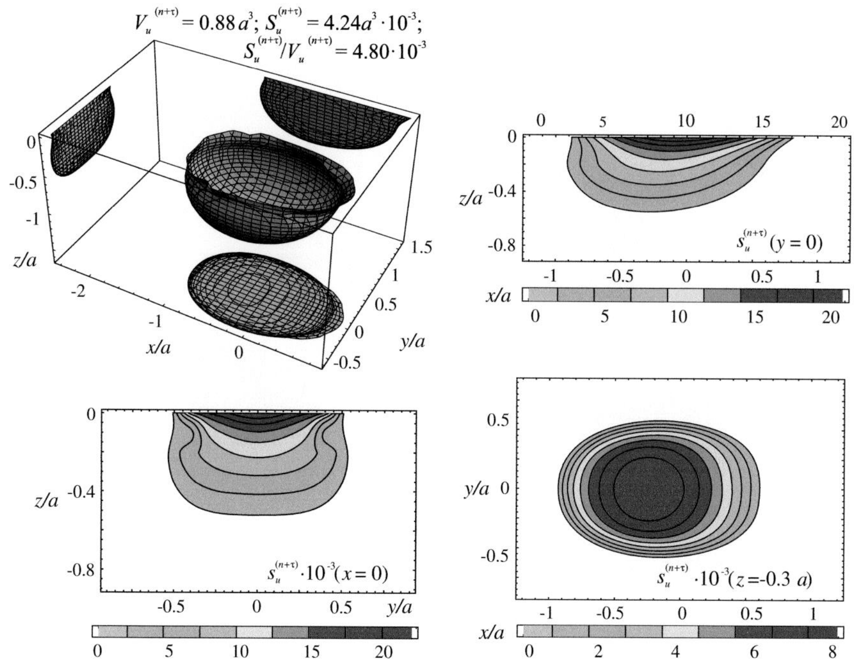

Fig. 3.2 Energy dangerous volume and its sections with specific entropy distributions for contact interaction with friction

Calculations of the three-dimensional stress–strain state in the neighborhood of the elliptic contact for $b/a = 0.574$ [4] show that the maximum value of the strain energy u is related to the maximum contact pressure $p_0^{(c)}$ in the following way:

$$u = \max_{dV}[u(F_c, dV)] = 0.47 p_0^{(c)}. \tag{3.28}$$

The limiting value of the strain energy $u^{(\text{lim})}$ under the action of the limiting force $F_c^{(\text{lim})}$ is

$$u^{(\text{lim})} = \max_{dV}\left[u\left(F_c^{(\text{lim})}, dV\right)\right] = 0.47 p_0^{(c,\text{lim})}. \tag{3.29}$$

In the calculations performed, maximum stresses σ_a due to noncontact bending in the contact area were the following

$$-0.34 \leq \sigma_a / p_0^{(c)} \leq 0.34.$$

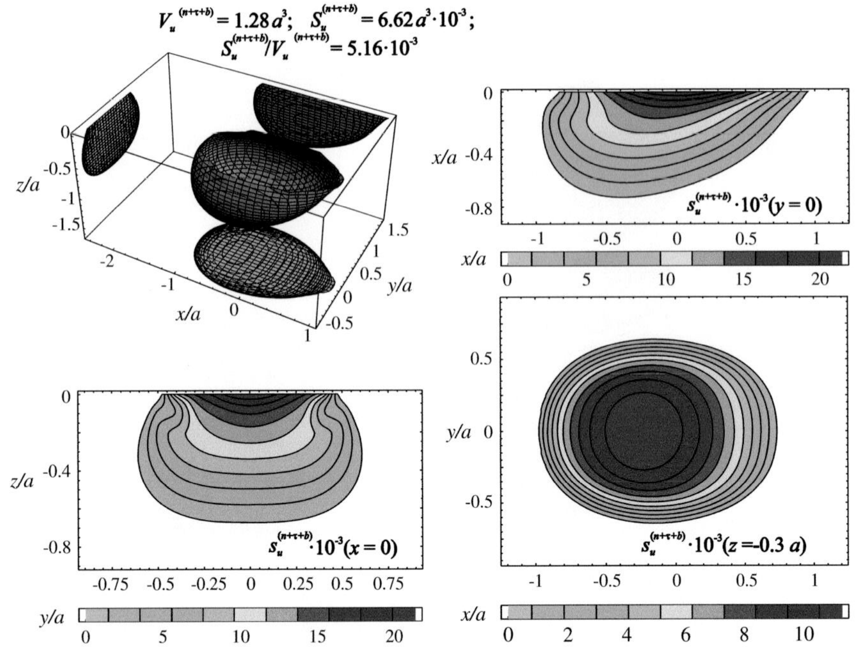

Fig. 3.3 Energy dangerous volume and its sections with specific entropy distributions for contact interaction with friction and tensile stresses $\sigma_a/p_0^{(c)} = 0.34$ in the contact area caused by noncontact bending

Tangential surface forces (friction force) are directed along the major semi-axis of the contact ellipse are:

$$p^{(\tau)}(x,y) = -fp^{(n)}(x,y) = -fp_0^{(c)}\sqrt{(1 - x^2/a^2 - x^2/b^2)}.$$

The specific entropy distribution calculated according to (3.26) shown in Figs. 3.1, 3.2, 3.3 and 3.4 can be considered to be the characteristic of the probability of appearance of local damages (initial cracks). The higher the specific entropy at a point of a dangerous volume, the greater is the probability of initiation of damage (crack) at this point. The values of dangerous volume and entropy are the integral damageability indices (including a possible number of cracks and their sizes) of a solid or a system.

From Figs. 3.1, 3.2, 3.3 and 3.4 for $p_0 = p_0^{(c)}$ and the friction coefficient $f = 0.2$ the maximum specific entropy is in the center of the contact area.

Under the joint action of contact pressure and tangential surface forces (friction) $s_u^{(n+\tau)}$, the maximum specific entropy increases by about 30 % in comparison with the maximum specific entropy $s_u^{(n)}$. The joint action of contact pressure, friction, and

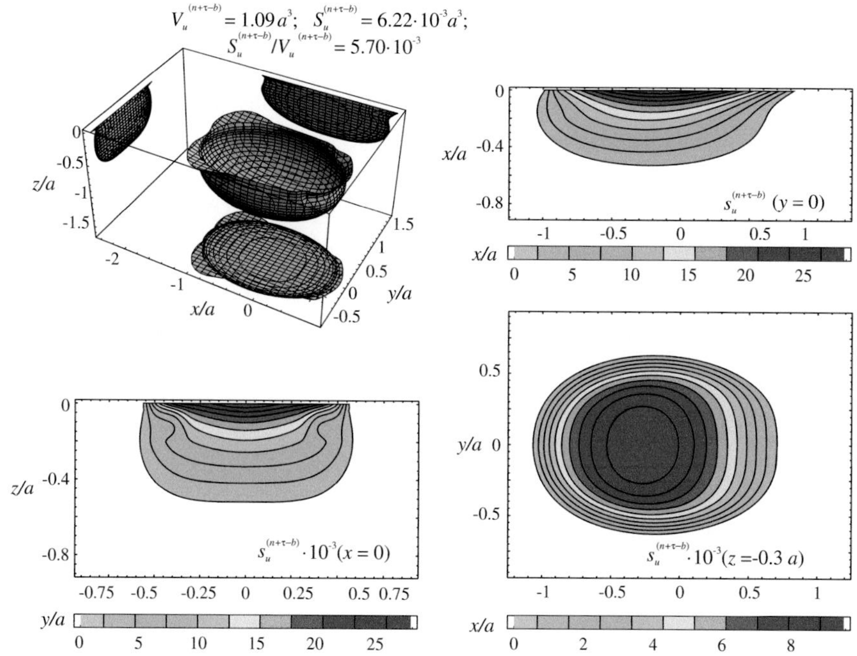

Fig. 3.4 Energy dangerous volume and its sections with specific entropy distributions for contact interaction with friction and compressive stresses $\sigma_a/p_0^{(c)} = -0.34$ in the contact area caused by noncontact bending

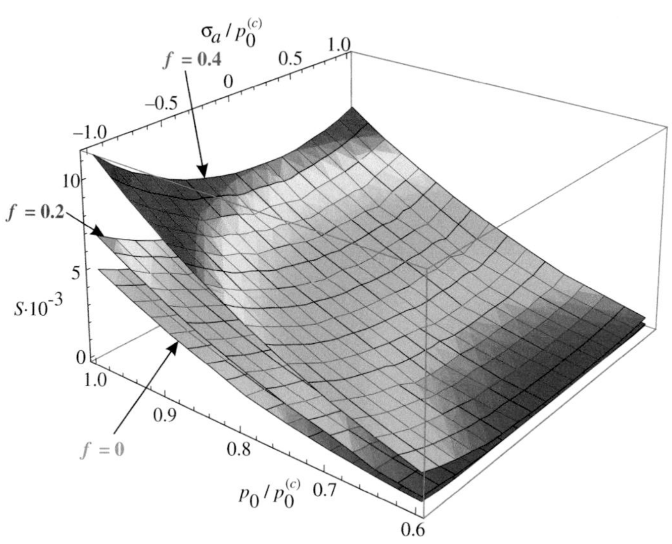

Fig. 3.5 Entropy versus contact and noncontact stresses

tension due to bending increases $s_u^{(n+\tau+b)}$ by about 30 % in comparison with $s_u^{(n)}$ At compressive bending, $s_U^{(n+\tau-b)}$ increases by about 60 % in comparison with $s_u^{(n)}$.

In case of frictional contact, the values of the dangerous volumes $V_u^{(n+\tau)}$, the entropy $S_u^{(n+\tau)}$, and the average entropy $S_u^{(n+\tau)}/V_u^{(n+\tau)}$ increase by about 6, 35 and 27 %, as compared to $V_u^{(n)}$, $S_u^{(n)}$, and $S_u^{(n)}/V_u^{(n)}$, respectively.

If both friction and tensile bending stresses are applied to the system, then the values of the dangerous volume $V_u^{(n+\tau+b)}$, the entropy $S_u^{(n+\tau+b)}$, and the average entropy $S_u^{(n+\tau+b)}/V_u^{(n+\tau+b)}$ are increased by approximately 54, 112, and 36 %, as compared to $V_u^{(n)}$, $S_u^{(n)}$, and $S_u^{(n)}/V_u^{(n)}$, respectively.

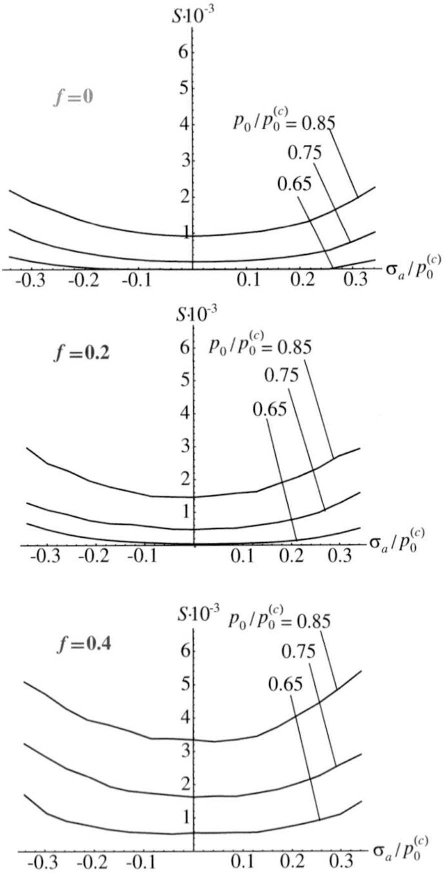

Fig. 3.6 Entropy versus the noncontact bending

If both friction and compressive bending stresses are applied to the system, then the values of the dangerous volume $V_u^{(n+\tau-b)}$, the entropy, $S_u^{(n+\tau-b)}$ and the average entropy $S_u^{(n+\tau-b)}/V_u^{(n+\tau-b)}$ increase by about 31, 98, and 50 %, as compared to $V_u^{(n)}$, $S_u^{(n)}$, and $S_u^{(n)}/V_u^{(n)}$, respectively.

A more detailed analysis of the considered effects might be done using Figs. 3.5, 3.6, and 3.7. They show a significant growth of entropy with increasing contact pressure, friction coefficient, and stresses caused by noncontact loads. The entropy increases almost at the same level for the same absolute values of tensile and compressive noncontact stresses. This effect may be due to the fact that the energy u calculated according to (3.23) attains positive values.

Fig. 3.7 Entropy versus contact stresses

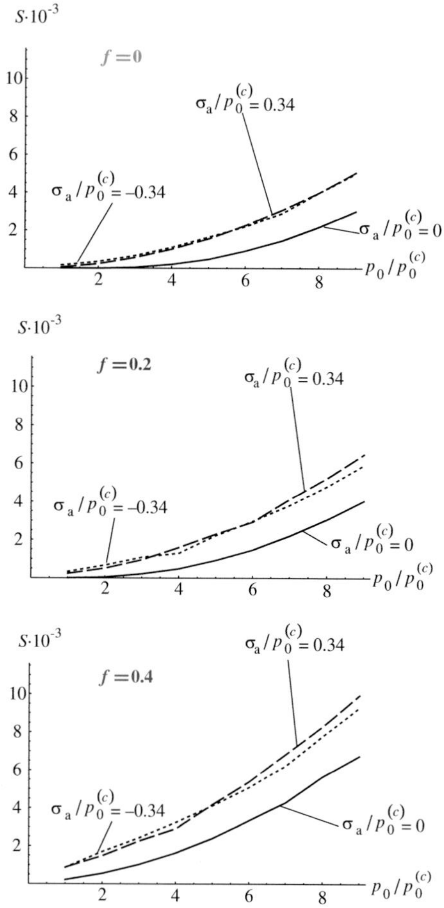

The main conclusion of Figs. 3.1, 3.2, 3.3, 3.4, 3.5, 3.6 and 3.7 is that not only friction, but also noncontact forces significantly change entropy characteristics in the neighborhood of the contact area.

Note that according to (3.18), (3.23)–(3.26), calculations were performed for the simplest case when the energy applied to the system is fully absorbed. Similar calculations may be done for effective energies u^{eff} determined by (2.53).

3.4 Interrelation of Motion, Damage, and Information

As shown above (Sect. 3.2) and in [17, 20, 24], the processes of irreversible damage in a mechanothermodynamic system generate Tribo-Fatigue entropy. This means that on the one hand, the interrelation of motion and damage should exist. On the other hand, it is obvious that the damage of a moving system alters its information state. Hence, there arises a *general problem of searching the interrelation*

$$\text{Mot} \Leftrightarrow \text{Inf} \Leftrightarrow \text{Dam}, \tag{3.30}$$

where Mot is the motion, Inf is the information, and *Dam* is the damage.

Specify problem (3.30) in the following order.

First, find the *function of interrelation of motion and information*, i.e.,

$$\text{Mot} \Leftrightarrow \text{Inf}. \tag{3.31}$$

The analysis is based on *Vavilov*'s functional

$$\log_2 P[0, X(0)] - \log_2 P[T, X(T)] = k \int_0^T \text{div} F(t, X, G) dt, \tag{3.32}$$

which specifies (3.31) to a first approximation, here $k = 1/\ln 2$ is the constant, $P(T, X(T))$ is the probability density function defining the position (state) X of a system at time T.

Let the system go from some initial state $X(0)$ to the finite state $X(T)$. This requirement is realized in the right part of functional (3.32). The left part of the functional consistently describes the information increment formed in the system for the time of its transition (from $t = 0$ to $t = T$), since $I_2 = -\log_2 P[T, X(T)]$ is the information, meaning that the system $F[t, X, G(t, X)]$ is in the state X at the time moment T, whereas $I_1 = -\log_2 P[0, X(0)]$ is the information, meaning that $X(0)$ was its initial state at the time moment $t = 0$. It is clear that $I_2 - I_1 = \Delta I$ is just the information increment. Note that the structure of expressions for definition of the information I_1 or I_2 is very complex. Suffice it to say that the initial values of the phase coordinates $X(0)$ are considered to be random and their distribution is characterized by the density $P(X, 0) = P_0(X)$.

The variable in the right hand-side of (3.32) depends on the choice of the form of the vector-control function $G = G(t, X)$ that enters as an argument the vector-function $F(t, X, G)$ present the right-hand side of (3.32). The latter describes the dynamic properties of an object of control or in other words its evolution (variation in time). Owing to this, $F(t, X, G)$ will be called simply the dynamic function; similarly, $G(t, X)$ will be called briefly the control function. Second, here, we are talking about the dynamics of an object not in the Eucledean (three-dimensional x, y, z) space but in the mathematical space; such a space is also called the phase space, or the space of states X. In (3.32), $X = (x_1, x_2, x_3, \ldots, x_n)^T$ is the vector of the system state in the n-dimensional space. If it is assigned, it is exactly known, at what point of the phase space the object is located. And, at last, under the integral in functional (3.32) there is not simply the dynamic function $F[t, X, G(t, X)]$, but its divergence $\mathrm{div}F(\cdot)$, i.e., the variation (divergence) of the vector field at a given point. The thing is: the vector-function $F[t, X, G(t, X)]$ can be expanded, as any vector, into components $f_1[t, X, G(t, X)], f_2[t, X, G(t, X)], \ldots, f_n[t, X, G(t, X)]$ that form the vector field of the control system together with the mathematical model of the control object

$$\dot{X} = F(t, X, G). \tag{3.33}$$

The divergence of the dynamic function $\mathrm{div}F(\cdot)$ can be interpreted as the degree and direction of divergence of two adjacent trajectories of closely located points;

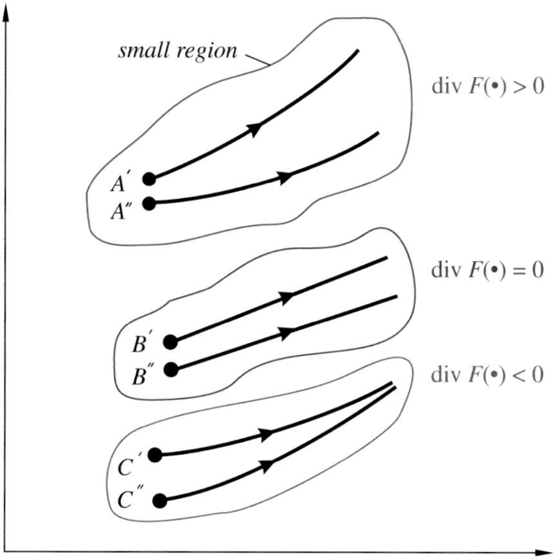

Fig. 3.8 Divergence as the characteristic of the divergence degree of phase trajectories

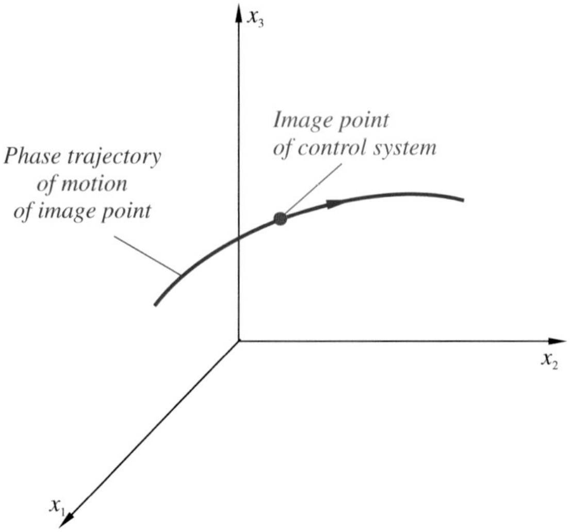

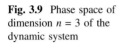

Fig. 3.9 Phase space of dimension $n = 3$ of the dynamic system

they are called the phase trajectories in some small region of the space of states X (Fig. 3.8). If trajectories diverge, the divergence is considered to be positive, if they converge—negative, and in the case when phase trajectories are parallel, then we are talking about zero divergence.

Let us say a few words about the vector-function F (t, X, G). It describes the dynamic properties of mathematical model (3.33) for the control object, i.e., the motion of its image point in the phase space of dimension n. Having chosen a particular form of the control vector G for the third-order control object, it is established that the motion of its image point occurs in the space of dimension $n = 3$, i.e., in the three-dimensional (x_1, x_2, x_3) space (Fig. 3.9).

In this case, $X = (x_1, x_2, x_3)^T$ is the control object state vector and $F(t, X, G) = (f_1(t, X, G), f_2(t, X, G), f_3(t, X, G))^T$ will be the vector-function. It should be said that mathematical model (3.33) of the control object is presented in its most general form. When practical problems are to be solved, it assumes a specific form.

Let us explain the aforesaid by the example of the economy. The vector-function $F(t, X, G)$ for the mathematical model of the distribution of investments between three branches will be of the form

$$F(t,X,G) = (-a_1x_1 + u_1, -a_2x_2 + u_2, -a_3x_3 + u_3)^T,$$

where a_1, a_2, a_3 are the coefficients. Mathematical model (3.33) in this case can be written in the form of the system of the three first-order differential equations:

$$\begin{cases} \dot{x}_1 = -a_1 x_1 + u_1; \\ \dot{x}_2 = -a_2 x_2 + u_2; \\ \dot{x}_3 = -a_3 x_3 + u_3. \end{cases}$$

Let us illustrate the strength and potentialities of the functional of type (3.32). Unfortunately, this can be done only for the simple case, namely: information transformations by noncontrolled linear stationary dynamic systems will be studied. General (very complex) mathematical model (3.33) for such systems becomes the most simple:

$$\dot{X} = BX. \tag{3.34}$$

Here, the $n \times n$ matrix is

$$B = \begin{bmatrix} b_{11} & b_{12} & \dots & b_{1n} \\ b_{21} & b_{22} & \dots & b_{2n} \\ \dots & \dots & \dots & \dots \\ b_{n1} & b_{n2} & \dots & b_{nn} \end{bmatrix},$$

and functional (3.32) reduces to the following expression;

$$\log_2 P(0, X(0)) - \log_2 P(T, X(T)) = k \int_0^T \operatorname{tr} B \, dt, \tag{3.35}$$

where spur tr of the matrix B is

$$\operatorname{tr} B = b_{11} + b_{22} + \dots + b_{nn}.$$

If in (3.35) it is assumed that $T = t$, then in spite of (3.34) the consideration can be supplemented by the *information function of the second-order dynamic system*

$$\Delta I = I_2 - I_1 = -k \cdot \operatorname{tr} B \cdot t. \tag{3.36}$$

This is what we have strived for: function (3.36) characterizes the information transformation in simplest dynamic system (3.34). In other words, the information function of dynamic system (3.36) is the first hint of the desired function of dialectic transformation (3.31). Indeed, function (3.36) allows one to qualitatively (in bites) define the information to be formed in the dynamic system (even the simplest one) when it goes from one state to another. In turn, this gives the possibility to set up a correspondence between dynamic processes and information that is generated by them. Thus, function (3.36) describes the *information state* of the second-order dynamic system and function (3.32) characterizes the *information state* of any systems.

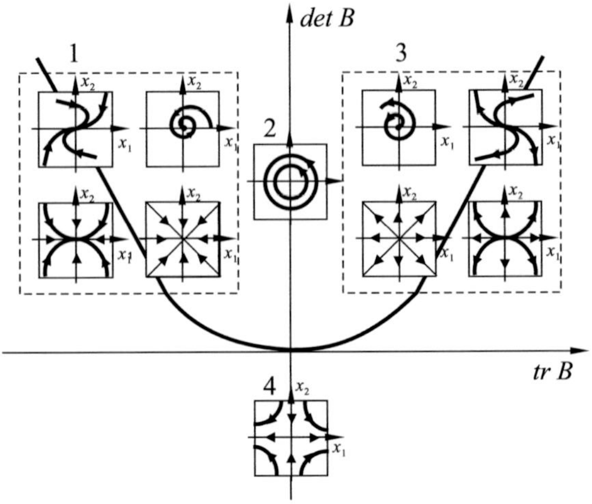

Fig. 3.10 Phase portraits of the second-order dynamic system

So, in the case of *linear stationary dynamic systems* (3.34), *information function* (3.36) is also linear.

The plot of this function is defined in terms of the dynamic properties of system (3.34) that in function (3.36) are given by the factor tr B. As a result of the analysis, it is a success to establish a correspondence between possible dynamic processes, which occur in linear stationary systems, and information function (3.36). Graphically, such correspondence is shown in Figs. 3.10, 3.11 and 3.12.

Figure 3.10 presents schematically (in frames) the typical possible phase "portraits" of system (3.34) when its dimensionality $n = 2$. They are plotted in the $x_1 - x_2$ coordinates in the form of the graphs of the phase trajectory along which the image

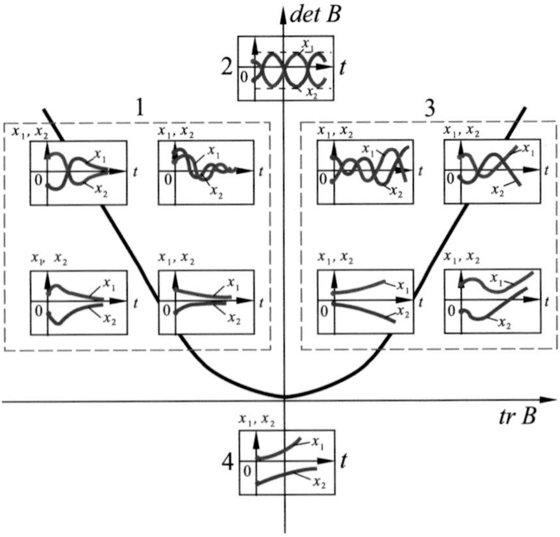

Fig. 3.11 Time plots of the second-order dynamic system

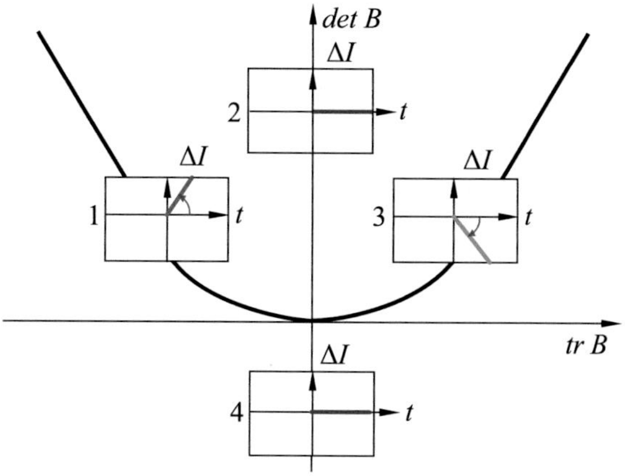

Fig. 3.12 Information increment due the system motion

point of the system moves. The arrows denote the motion directions. The entire set of the phase portraits (Fig. 3.10, ten portraits) is located in the common plane, whose abscissa axis is the trace of the matrix B (tr B), and the ordinate axis is its determinant det B. The graph of the function det B = (tr B)^2 .can be easily plotted. It is the parabola in Fig. 3.10. Then the phase portraits of the second-order dynamic system are located relative to this parabola, as shown in Fig. 3.10. It is seen that four groups (1, 2, 3, 4) of phase portraits are available, each of which lies in the characteristic zone of the plane depending on det B and tr B ratios and signs.

Figure 3.11 shows the graphical representation of the analyzed system, but in another form and also the typical plots of motion (in rectangular frames) determined by the x_1 and x_2 coordinates in time t.

These plots are consistent with the phase portraits in Fig. 3.10 (the notation of groups 1, 2, 3, and 4 of the plots in the both figures is the same). In Fig. 3.11: 1—oscillation ional and asymptotic converging processes, 2—sustained oscillation ional processes, 3—oscillation ional and asymptotic diverging processes, 4—unstable processes.

That shown in Figs. 3.10 and 3.11 is long and well known for specialists; it is the usual result of investigation of the dynamics of simplest system (3.34). Figure 3.12 has a new interpretation of the analyzed data: here the plots of information function (3.36) in the coordinates of the information increment ΔI—the time t. Figure 3.12 gives the same notations of groups 1, 2, 3, and 4 as in Figs. 3.10 and 3.11. The phase portraits (Fig. 3.10) and the time functions (Fig. 3.11) of system (3.34) are consistent with the plots of information function (3.36) in the coordinates of the information increment ΔI—the time t. Figure 3.12 displays the same notations of groups 1, 2, 3, and 4 as in Figs. 3.10 and 3.11.

It is seen that only plot 1 (Fig. 3.12) of the information increment in time is consistent with four phase portraits, which compose group 1 in Fig. 3.10, and, hence, with four time functions, which compose group 1 in Fig. 3.11. Similarly, only plot 3 (Fig. 3.12) of the information increment in time is consistent with four phase portraits, which compose group 3 in Fig. 3.10, and, hence, with four time functions, which compose group 3 in Fig. 3.11 (Fig. 3.12). In both these cases (Plots 1 and 3 in Fig. 3.12), the same linear information function is revealed, but with the following substantial feature: information increment will be either positive—at converging (oscillation ional and asymptotic) processes (Group 1 in Figs. 3.10 and 3.11), or negative—at diverging (oscillation ional and asymptotic) processes (Group 3 in Figs. 3.10 and 3.11). Stable (by Lyapunov) dynamic processes (Group 2 in Figs. 3.10 and 3.11) and unstable processes (Group 4 in Figs. 3.10 and 3.11) do not produce new information at all (Plots 2 and 4 of zero inormation increment in Fig. 3.12).

Thus, the information increment can be negative or positive.

Now, mention one more conclusion that directly follows from the plots in Figs. 3.8, 3.9, 3.10, 3.11 and 3.12, but it can be obtained from a further analysis of the obtained solution: linear information functions shown in Fig. 3.12 also remain the same for dynamic systems of higher than second order. This means that these—information functions (Fig. 3.12) are more general in character than phase portraits (Fig. 3.10) or time diagrams (Fig. 3.11).

It has been found above that dynamic systems in some quite certain state are unable in general to generate new information (Groups 2 and 4 in Figs. 3.10, 3.11 and 3.12). The fact that at regular non-damping oscillation ions (Plots 2 in Figs. 3.10, 3.11 and 3.12), new information is not generated in the system. Naturally, it is clear: a sad monotony of such repeating oscillation ions can generate nothing since it is a reversible process. But it is the fact that definite unstable systems (Plots 4 in Figs. 3.10, 3.11 and 3.12) do not also generate any new information. This fact is unclear because it is a question of the irreversible process.

The results obtained can be interpreted from the viewpoint of the theory of damageability of moving objects.

As known, the theory of damage of deformable solids has been developed successfully and long since. But the theory of damageability of moving objects seems to be unavailable. First, attempts are made to formulate, at least, a general approach to the construction of such a theory.

Again, the most general and ordinary concept will yield that any object (any system) evolves in time and any process of their development (existence) ends with death. Reasons for death are extremely diverse and complex: from nuclear reactions and cosmic interactions (Universe's objects) to natural aging (living organisms). Generalizing the aforesaid, it can be said: these reasons are damages, whose nature and types are also diverse and complex. Hence, a conclusion can be made: the most important general property of each system is its ability to accumulate damages at all possible conditions of existence, for example, at deformation, motion, relative rest, etc., i.e., at inevitable interactions with other bodies and systems, the environment and flows of any nature particles, etc. It is natural that the process of damage accumulation must be different depending both on the properties of an evolving system itself and the properties of flow substances, etc.

Now, for (3.30) to be defined concretely, let us study the damageability of solids in time. We consider that the main types of kinetic processes of damage $\omega_{\Sigma t}$ of an object (solid, Tribo-Fatigue system) can be described by the simplest power equation [14, 20]

$$\omega_{\Sigma t} = \left[1 - \left(1 - \frac{t}{T_{\otimes}}\right)^{h}\right]^{q}, \tag{3.37}$$

where $T_{\otimes}$ is the life (longevity); $h \geq 1$, $q \geq 1$ are the controlling parameters. According to (3.37) if $h \geq 1$, $q = 1$, then the phenomena of material softening (the convex curve in Fig. 3.13) are dominant, while, on the contrary, at $q > 1$, $h = 1$ these are the phenomena of material hardening (the concave curve in Fig. 3.13). It is not difficult to understand that these phenomena are caused by the dynamics of motion: the strengthening is "realized" on the left branch of the parabola (Group of plots 1 in Fig. 3.11) when $\mathrm{div}F(\cdot) < 0$ (Fig. 3.8) and the softening—on the right branch of the parabola (Group of plots 3 in Fig. 3.11) when $\mathrm{div}F(\cdot) > 0$ (Fig. 3.8). At $h = 1$, $q = 1$ the system is stable (dotted line in Fig. 3.13 and curves 2 in Fig. 3.11), so that $\mathrm{div}F(\cdot) = 0$ (Fig. 3.8). In the general case, $h > 1$, $q > 1$ the processes of hardening–softening of the system are determined by the parameter ratio h/q and are described by more complex (S-shaped) curves (Fig. 3.13) that are consistent with curves 4 in Fig. 3.11.

For any fixed time moment, let us introduce a unique characteristic of systems—damageability index

$$\omega_j = \omega_{st} - \omega_{\Sigma}, \tag{3.38}$$

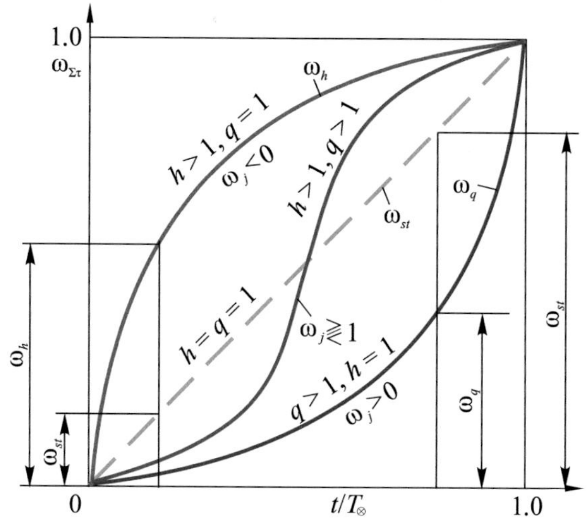

Fig. 3.13 Schemes of possible kinetic processes of irreversible damages

where $\omega_\Sigma = \omega_h$ or $\omega_\Sigma = \omega_q$ is the damageability level of a real system, and ω_{st} is the damage level of some ideal system corresponding to that of the real system.

Then it turns out that values (3.37) can belong to three characteristic classes: $\omega_j > 0$; $\omega_j < 0$ and $\omega_j = 0$ (Fig. 3.13). There arc the same three characteristic classes for the information increment ΔI (Fig. 3.12). This means the existence of a relationship between ω_j and ΔI. By summing up the damageability index in time, Eq. (3.36) may be continued in the following way:

$$\Delta I(t) = -k \cdot \mathrm{tr}\, B \cdot t = a_S\, \Sigma\omega_j(t), \tag{3.39}$$

where a_S is the transition function between ΔI and $\Sigma\omega_j$.

This is the solution of problem (3.31).

Now, let us present the hardening and softening functions and the damageability indices on a single field *det B–tr B* (Fig. 3.14).

The analysis of the data plotted in Figs. 3.10, 3.11, 3.12, 3.13 and 3.14 yields three general conclusions.

First, the left branch of the parabola plotted in the *det B–tr B* coordinates illustrates the stable dynamic processes that generate a positive linear information function due to the development of nonlinear hardening of an object.

Second, the right branch of the parabola plotted in the *det B–tr B* coordinates corresponds to the unstable dynamic processes that generate a negative linear information function due to the development of nonlinear softening of an object.

Third, the vertex of the parabola plotted in the *det B–tr B* coordinates can be consistent with the radically different states of the system: (a) self-sustained oscillatory processes (above the parabola vertex—on the *det B* axis)—new information is not generated here since the damageability index $\omega_j = 0$; (b) unstable processes (below the parabola vertex—on the same *det B* axis), when a zero information function is again generated, and the damageability index $\omega_j \gtreqless 0$. To understand this contradiction, one needs to search some specific features of this

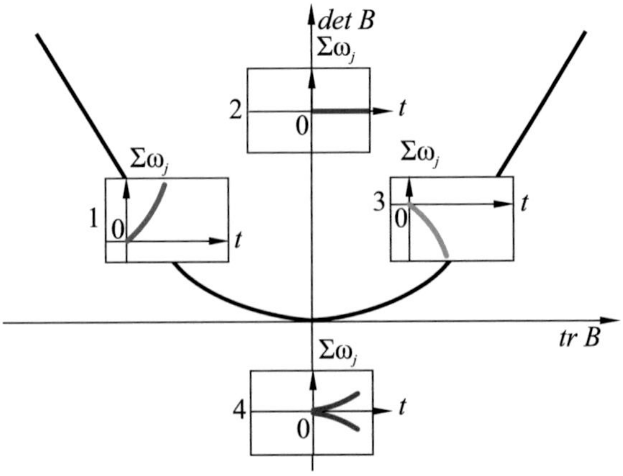

Fig. 3.14 Change in the damageability index due to the system motion

Table 3.2 Interrelation of motion, information, and damageability

Group (Figs. 3.11 and 3.12)	Signs		Processes		Systems	Damageability		Information functions
		Tr B	Time	Physical		Processes	Index	
1	+	−	Converging (oscillatory and asymptotic)	Irreversible	Dissipative, div $F(\cdot) < 0$	Hardening	$\omega_j > 0$	Linear positive
2	+	0	Oscillatory continuous	Reversible	Conservative, div $F(\cdot) = 0$	Stable	$\omega_j = 0$	Zero
3	+	+	Diverging (oscillatory and asymptotic)	Irreversible	Non-conservative (dissipative), div $F(\cdot) > 0$	Softening	$\omega_j < 0$	Linear negative
4	−	0	Unstable	Irreversible	Conservative, div $F(\cdot) = 0$	Stable, hardening–softening fluctuations	$\omega_j \gtreqless 0$	Zero

instability that are fundamentally different from the instability of oscillatory and asymptotic diverging processes.

Our available basic data on second-order dynamic systems are systematized in Table 3.2. Let us add: from the energy point of view, systems are dissipative and conservative. For dissipative systems, the divergence div $F(t, X, C) \neq 0$ and for conservative systems, div $F(t, X, C) = 0$.

The analysis of Table 3.2 allows us to answer the above question: why does not any information arise as a result of an unstable process? It appears because it is the case of a conservative system, for which the divergence div $F(t, X, C) = 0$ and $\omega_j = 0$. Hence, the motion does not initiate information if the divergence and so the damageability of the system are zero. In other words: a conservative system cannot produce new information.

Let us comment the development of hardening-softening processes. It follows from the analysis of Eq. (3.37) at $q > 1$, $h > 1$ (S-shaped curve in Fig. 3.13). Systems exist, in which the softening process in time at an assigned load cannot be spontaneously replaced by the hardening process; these systems are called softening.

Only reverse spontaneous hardening is always replaced in time by spontaneous softening. Hence, the system can inevitably reach critical and supercritical states—up to its disintegration ($\omega_\Sigma \to \infty$).

Thus, it is possible that for some systems there is a prohibition of the sequence of spontaneous processes of softening → hardening or in other words the rule of priority is obeyed: hardening → softening. If it is so, then taking into account the data of Table 3.2 and Eq. (130), it should be stated: the dissipative system, for which div $F(t, X, U) < 0$, can with time spontaneously go into the conservative state, for which div $F(t, X, U) > 0$, but the inverse is impossible. This means that there exists (at least, for softening systems) a prohibition of the conversion sequence of negative information into positive; only the conversion sequence of positive information into negative is permitted.

Thus, the rule of priority corresponds to the idea, according to which any system inevitably dies; in this case, must its information function become "extremely negative?" or zero? Up to now there is no answer to this question.

One more remark should be made. One must be clearly aware of the fact that hardening–softening is a twofold form of the fundamental process: damages in time. According to (3.37), just the hardening-softening process ratio (h/q) controls the kinetics of damages.

The interrelation of motion, information, and damageability (Fig. 3.15) is found from the simplest example (Figs. 3.11, 3.12, 3.14 and Table 3.2). Its analysis permits one to make the following basic conclusions: *the motion generates new information in the system if its damageability index is not equal to zero; the information is positive, when the system is hardened, and is negative, when it is softened. It is the reciprocity relation, or the Tribo-Fatigue triada.*

Generalize the aforesaid. *Matter as an objective reality given us in sensations and measurements, its motion and damage are the information sources in nature. Forces (energy) and Λ-interactions (particles, bodies, systems) are the reasons for*

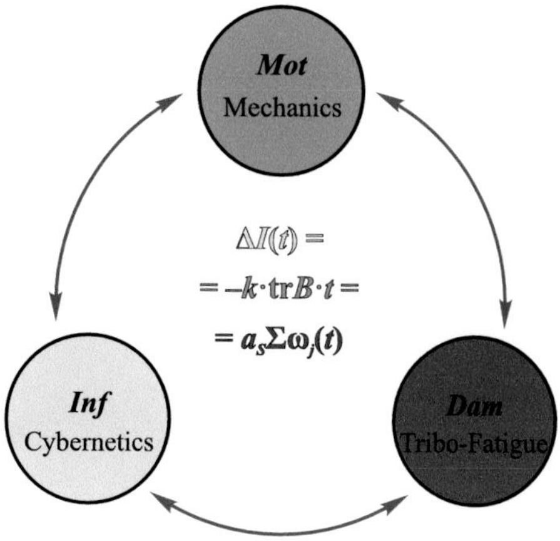

Fig. 3.15 Interrelation of motion, information, and damage (Tribo-Fatigue triada)

motion and damage of matter and any objects composed of matter. Forces (energy) and Λ-interactions hence cause the generation and accumulation of information. What is this process? How is the process completed? The first answers to these difficult questions are given in [20, 23, 25] applied to peculiar—living and intelligent—systems.

3.5 Evolution of a System and Entropy

It has been stated above and in [20, 22, 24] that the mechanothermodynamic state of a system can be in principle characterized by total changes in the dissipated and absorbed parts of energy, or entropy.

Here, let us mention one important problem. In [16, 18, 19], it is shown that the damages due to different-nature loads (e.g., thermal and mechanical) do not exhibit the additivity properties; on the contrary, they have the property of nonlinear interaction. The results on such interactions are described by the functions $\Lambda_{\sigma/p}$, $\Lambda_{T/M}$ (see Chap. 2). In [16, 24], the Λ-functions were defined concretely for some deformable systems (see Chap. 2).

Let us support the aforesaid experimentally.

Comparative experiments were performed on the damage of a deformable system at rolling friction and complex loading: rolling friction + mechanical fatigue. In the both cases, during tests the contact pressure p_0 was increased stepwise over its range (Fig. 3.16, stages I, II, …, XII). When testing the shaft/roller system the approach δ_c of the axes of this pair of the elements was measured under the rolling friction conditions (when the amplitude of cyclic stresses $\sigma_a = 0$) and under mechano-rolling fatigue conditions (at $\sigma_a = 0.8_{\sigma-1}$ and $\sigma_a = 1.0_{\sigma-1}$ where $\sigma - 1$ is the fatigue limit). From Fig. 3.16, it is seen that the process of accumulation of complex wear-fatigue damages essentially slows down in comparison with

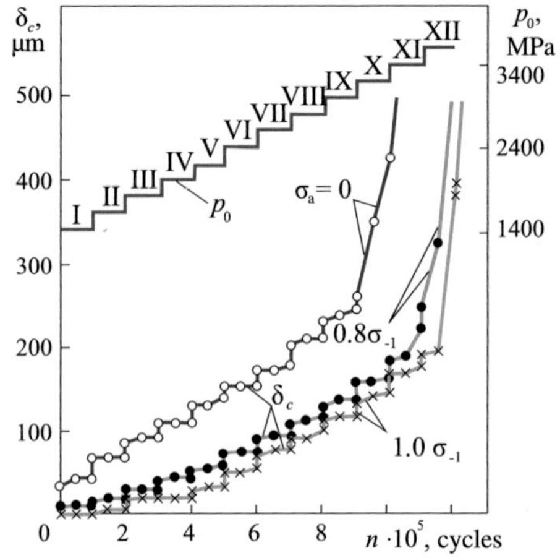

Fig. 3.16 Test results

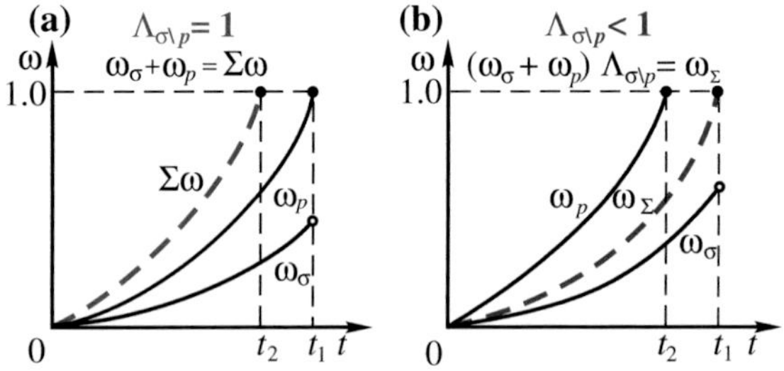

Fig. 3.17 Scheme of summation **a** and interaction **b** of damages

damages at rolling friction. In this case, the range of contact pressure corresponding to normal friction widens by about 14 %. Based on these experimental data, let us discuss the difference between the processes of summation and interaction of damages, considering that the limiting state is reached when the damage measure ω becomes critical, i.e., ω = 1.

Let damages due to contact (ω_p) and noncontact (ω_σ) loads be accumulated for the time t_1 in the manner shown in Fig. 3.17, a: the critical state is achieved by none of these criteria ($\omega_p \ll 1.0$; $\omega_\sigma \ll 1.0$). If the damages are summed up $(\omega_p + \omega_\sigma) = \Sigma\omega$, then in the case of wear-fatigue tests, the critical state ($\Sigma\omega = 1.0$) will be reached for the time $t_2 < t_1$. However, such a prediction appears to be apparently incorrect, as applied to the experimental data in Fig. 3.16. If it is taken into account that the damages caused by contact and noncontact loads interact

$$(\omega_p + \omega_s)\Lambda_{\sigma\backslash p} = \omega_\Sigma, \tag{3.40}$$

so that under the analyzed conditions the interaction function $\Lambda_{\sigma/p} < 1$, then the scheme adequately illustrating the experimental data in Fig. 3.16 looks like the one shown in Fig. 3.17b, illustrating that at rolling friction the critical state is reached by ω_p for the time t_2, whereas at mechanical fatigue (ω_σ) it does not occur even for

$t_1 \gg t_2$. Under the conditions of wear-fatigue tests characterized by the damage measure ω_Σ the life (t_1) turns out to be larger than the one at rolling friction (t_2).

Since the irreversible damageability is the function of the effective energy absorbed in the system (expression (3.5)), from (3.40) the general conclusion can be made using Figs. 3.16 and 3.17: at wear-fatigue damage effective energies due to contact (U_p^{eff}) and noncontact (U_σ^{eff}) loads are not summed up, they interact dialectically. Now, the principle of interaction of the effective energy components in the Tribo-Fatigue system [16, 18–20] can be written:

$$\left(U_\sigma^{eff} + U_p^{eff}\right)\Lambda\left(\omega_\sigma \leftrightarrows \omega_p\right) = U_\Sigma^{eff}, \quad \Lambda_{\sigma\backslash p} \gtreqless 1. \tag{3.41}$$

According to (3.41), the result (U_Σ^{eff}) of the interaction of damages ($\omega_\sigma \leftrightarrows \omega_p$) and, hence, of energies is governed both by the loading conditions and by the direction of the processes of hardening–softening ($\Lambda \gtreqless 1$) [16, 18, 19]. From (3.41), it follows that at $\Lambda(\omega_\sigma \leftrightarrows \omega_p) = 1$, the particular case of interaction of effective energies (and, hence, of damages)—their summation—is possible.

Developing the ideas of interaction of effective different-nature energies according to model (3.41) [4, 16, 18, 19, 24] yields a diversity of new conclusions since it involves a physically clear result: real damage and failure of systems. Four first surprises of Tribo-Fatigue ([24]) could be neither understood, nor described without the knowledge of principle (3.41).

According to the available data [5, 10], when the behavior of thermodynamic systems was analyzed, the problem of interaction of the dissipated part of different-nature energies was not stated. It is natural that a possible interaction of entropy produced by mechanothermodynamic forces and flow has not been investigated when different irreversible processes (3.3), (3.11), and (3.15) are realized. But since the entropy and energy relation $S(U)$ is fundamental (3.1) and (3.2), resting upon the abovesaid in the general case of the analysis of the mechanothermodynamic state of systems, *sum* (3.11) *of thermodynamic and Tribo-Fatigue entropies should be written having regard to possible Λ-interactions*:

$$S_{total}(t) = \left(S_T(t) + S_{TF}(t)\right)\Lambda_{T\backslash TF}. \tag{3.42}$$

As mentioned, the fundamentals of the theory of Λ-interactions of irreversible damages in Tribo-Fatigue systems have been formulated and to some extent

Table 3.3 Parameters for the mechanothermodynamic state of different systems

Parameter	Characteristic
$\text{div}\, F(\cdot) \gtreqless 0$	Relative motion of physical points of matter or elements of a system (converging, diverging and other processes)
$\omega_j \gtreqless 0$	Nature of irreversible damageability (hardening, softening, etc.)
$\Lambda \gtreqless 1$	Direction and intensity of interaction of irreversible any nature damages
$\pm\Delta I$	Information changes in the process of system motion and damage

developed up to now [4, 16, 18, 19, 22, 24]. Creating the theory of irreversible Λ-interactions in mechanothermodynamic systems is waiting for its researcher. But Eq. (3.42) in combination with the results reported in [17, 24] enables one to analyze, in principle, the mechanothermodynamic state of systems.

To analyze (3.42), let us use four parameters which have been analyzed above (introduced in [20, 22, 24]) and given in Table 3.3.

It is now possible to plot, for example, Fig. 3.18. For the definite parameter ratios presented in Table 3.3, Eq. (3.42) predicts various and complex "trajectories" of entropy.

In the course of evolution, the system can go, for example, into stable and equilibrium states and come out from them as many times as possible under the particular conditions of its existence; the observed points A_1, A_2 of the system can approach each other and come apart or move practically parallel [22, 24]; the system can undergo bifurcations and other (more complex) transformations. Figure 3.18b illustrates that bifurcations are peculiar to softening systems with the

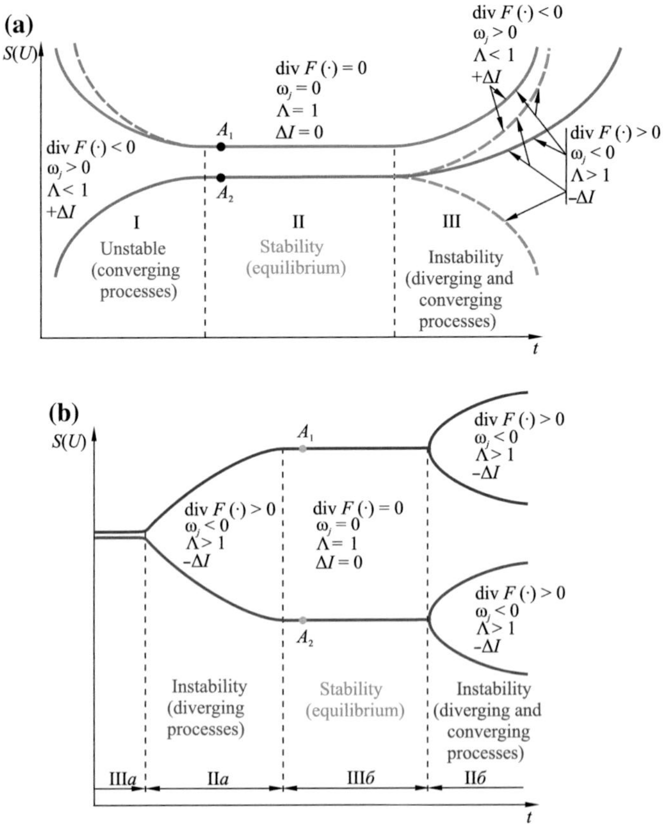

Fig. 3.18 Possible transitions of the system from unstable to stable state and, vice versa (**a**) and the onset of bifurcations (**b**)

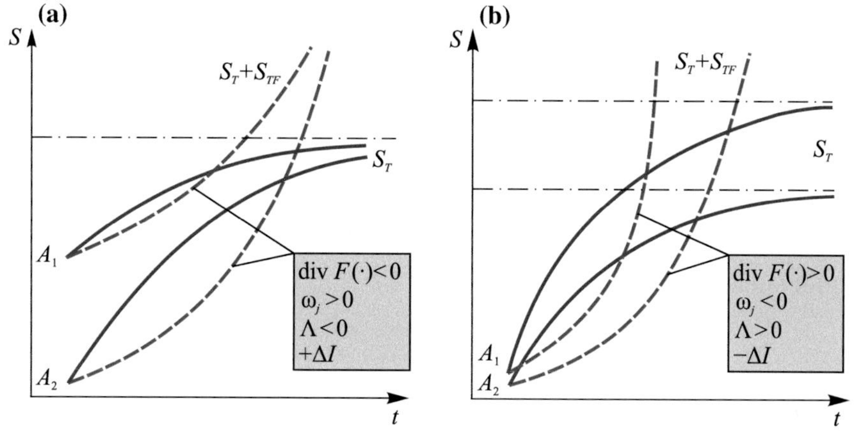

Fig. 3.19 Evolution of the thermodynamic (S_T) or the mechanothermodynamic ($S_T + S_{TF}$) state of the system (A_1, A_2): **a** oscillatory and asymptotic converging processes; **b** oscillatory and asymptotic diverging processes

negative information function. It is then natural that the question arises: what is the difference between the mechanothermodynamic and thermodynamic processes?

An answer to this question is illustrated by Fig. 3.19. Here the solid lines stand for the predicted behavior of the thermodynamic system, for which in (3.42) it is assumed that $S_{TF} = 0$ and $\Lambda_{T/TF} = 1$; let the entropy S_T of such a system tend to some (for example, local) maximum. The behavior of the mechanothermodynamic system is shown by the dotted lines in Fig. 3.19, assuming that in (3.42) $S_T \neq 0$ and $\Lambda_{T/TF} > 1$. The initial state of the both systems is assumed to be identical (points A_1, A_2). The fate of the system in the both cases is determined by the intensity of numerous irreversible internal processes caused by a diversity of reasons. But it will be fundamentally different for comparable systems.

On the other hand, the trajectory of the mechanothermodynamic state ($S_T + S_{TF}$) cannot coincide with that of the thermodynamic state (S_T) since in the first case, there appears a nonzero addition of Tribo-Fatigue entropy ($S_{TF} > 0$). This motivates quantitative differences in the trajectories of the systems to be compared. On the other hand, the principal difference is seen in their behavior: when the entropy of the thermodynamic system attains, for example, a local maximum (equilibrium state), the mechanothermodynamic system can have no such maximum—it will be in the nonequilibrium state. This is observed in the cases of the converging (Fig. 3.19a) and diverging (Fig. 3.19b) processes (Fig. 3.11), as well as of hardening and softening of systems in time, in which new positive or negative information is generated (Fig. 3.12). It should be noted the conservative mechanical system (Table 3.2) is in essence identical to the thermodynamic system since for it $S_{TF} = 0$. Some generalizations regarding a comparative behavior of thermodynamic and mechanothermodynamic systems are available in [21].

Thus, the fundamentals of the general theory of evolution of MTD systems are presented here using the entropy concepts. In Sect. 3.1, the theory of evolution of the same systems is developed proceeding from the energy concepts. Now the task is to make a summary analysis of evolution of MTD systems. This must be done since figuratively speaking, energy and entropy are two sides of the same coin—the damageability of the MTD system.

References

1. Amiri, M., Khonsari, M.M.: On the thermodynamics of friction and wear—a review. Entropy **12**, 1021–1049 (2010)
2. Bryant, M.D.: Entropy and dissipative processes of friction and wear. Trans. Fac. Mech. Eng., Belgrade, 55–60 (2009)
3. Doelling, K.L., et al.: An experimental study of the correlation between wear and entropy flow in machinery components. J. Appl. Phys. **88**, 2999–3003 (2000)
4. Feynman, R.: Feynman's Lectures on Physics. Mir, Moscow (1963). (in Russian)
5. Kondepudi, D., Prigogine, I.: Modern Thermodynamics: From Heat Engines to Dissipative Structures. John Wiley & Sons, Chichester (1998)
6. Mase, G.: Theory and Problems of Continuum Mechanics. McGraw-Hill, New York (1970)
7. Naderi, M., Khonsari, M.M.: An experimental approach to low-cycle fatigue damage based on thermodynamic entropy. Int. J. Solids Struct. **4**, 875–880 (2010)
8. Naderi, M., Khonsari, M.M., Amiri, M.: On the thermodynamic entropy of fatigue fracture. Proc. Royal Soc. Ser. A **466**, 423–438 (2010)
9. Naderi, M., Khonsari, M.M.: A thermodynamic approach to fatigue damage accumulation under variable loading. J.Mater. Sci. Eng. Ser. A **527**, 6133–6139 (2010)
10. Nicolis, G., Prigogine, I.: Exploring Complexity (An Introduction). Freeman & Company, New York (2003)
11. Prokhorov A.M. (ed.): Physical Encyclopaedic Dictionary. Soviet Encyclopaedia, Moscow (1983) (in Russian)
12. Sedov, L.I.: Mechanics of a Continuous Medium. Nauka, Moscow (1973). (in Russian)
13. Sherbakov, S.S., Sosnovskiy, L.A.: Mechanics of Tribo-Fatigue Systems. BSU Press, Minsk (2010). (in Russian)
14. Sosnovskiy, L.A., et al.: Reliability. Risk. Quality. BelSUT Press, Gomel (2012). (in Russian)
15. Sosnovskiy, L.A., Sherbakov, S.S.: Mechanothermodynamic system and its behavior. Continuum Mech. Thermodyn. **24**(3), 239–256 (2012)
16. Sosnovskiy, L.A.: Fundamentals of Tribo-Fatigue. BelSUT Press, Gomel (2003). (in Russian)
17. Sosnovskiy, L.A.: Mechanics of the irreversible damages caused by contact and noncontact load. In: Proceedings of World Tribology Congress III, Washington, 12–16 Sept 2005, pp. 2–8 (2005)
18. Sosnovskiy, L.A.: Tribo-Fatigue. Wear-Fatigue Damage and Its Prediction (Foundations of Engineering Mechanics). Springer, Berlin (2005)
19. 索斯洛夫斯基著, L.A.: 摩擦疲劳学 磨损—疲劳损伤及其预测. 高万振译—中国矿业大学出版社 (2013) (in Chinese)
20. Sosnovskiy, L.A.: L-Risk (Mechanothermodynamics of Irreversible Damages). BelSUT Press, Gomel (2004). (in Russian)
21. Sosnovskiy, L.A., Makhutov, N.A.: Reiability of Tribo-Fatigue systems in the non-classical statement. In: Conference on Sirvivability and Construction Material Science, 22–24 Oct 2012, Moscow, p. 53. Press of the A.A. Blagonravov Institute of Mechanical Engineering, Moscow (2012)

22. Sosnovskiy, L.A., Sherbakov, S.S.: Surpises of Tribo-Fatigue. BelSUT Press, Gomel (2005). (in Russian)
23. Sosnovskiy, L.A.: Intelligent dynamic systems: problem and research perspectives. mechanics 2011. In: Proceedings of V Belarusian Congress on Theoretical and Applied Mechanics, vol. 1, pp. 64–79. Press of the Joint Institute of Mechanical Engineering of NAS of Belarus, Minsk (2011) (in Russian)
24. Sosnovskiy, L.A.: Mechanics of Wear-Fatigue Damage. BelSUT Press, Gomel (2007). (in Russian)
25. Sosnovskiy, L.A.: On dynamic systems with elements of intelligence. In: Zhuravkov M.A., et al. (eds.) Proceedings of VI International Symposium on Tribo-Fatigue, 25 Oct–1 Nov 2010, Part 2, pp. 573–582. BSU Press, Minsk (2010) (in Russian)
26. Sosnovskiy, L.A.: Statistical Mechanics of Fatigue Damage. Nauka i Tekhnika, Minsk (1987). (in Russian)

Chapter 4
Principles of Mechanothermodynamics

Abstract Four principles of mechanothermodynamics (MTD) are formulated and stated. The first principle establishes the generalized law of damageability, the second—its main cause, the third—its scale, and the fourth—the interrelation of motion, damage, and information. In total, these fundamental principles are a basis for the generalized theory of A-evolution of inorganic and organic systems.

4.1 General Notions

The results presented in Chaps. 2 and 3 with regard to [13, 18, 21, 24, 27, etc.] are generalized in Fig. 4.1. It can be seen that the state of a system can be equivalently described in terms of either energy or entropy. The main drawback of such descriptions is the known unreality of energy and, hence, of entropy: physical energy carriers are not detected and, apparently, do not exist. They cannot be touched, as *Feynmann* [3] said figuratively. Damages are an entirely different matter: they are physically real, can be touched, and in reality define any of the conceivable states of material bodies and systems. The kinetic process of their accumulation, as well as the time steam is inevitable and unidirectional. If mechanothermodynamics considers the damage of a system as its fundamental physical property (and duty), it can be shown that based on it, the consistent general theory of evolution of any systems, including living and intelligent, can be created. For instance, the idea of life as of a special method of damage accumulation (biological, mechanical, intellectual, etc.) is developed in [21, 25].

Thus, the attempt was made above (see Chaps. 2 and 3) to formulate the basic tenets of a new (or, better to say, integrated) physical discipline—mechanothermodynamics with the use of the energy principles. This discipline combines two branches of physics in an effort not to argue or not to compete with each other, but to take a fresh look at the world and its evolution (Fig. 4.2).

Figure 4.3 shows that the *principles of mechanothermodynamics can be formulated in two ways*: (1) *mechanics → Tribo-Fatigue → mechanothermodynamics*

L. Sosnovskiy and S. Sherbakov, *Mechanothermodynamics*,
DOI 10.1007/978-3-319-24981-0_4

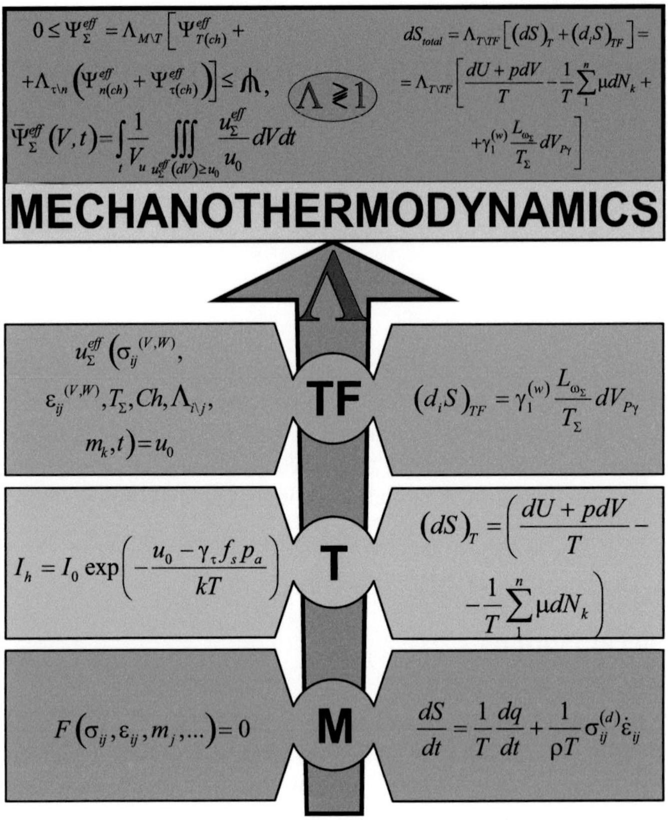

Fig. 4.1 Energy (*left*) and entropy (*right*) approaches to the development of mechanothermodynamics (*M*—mechanics, *T*—thermodynamics, *TF*—Tribo-Fatigue)

and (2) *thermodynamics* → *Tribo-Fatigue* → *mechanothermodynamics*. Thus, *Tribo-Fatigue has become a bridge for transition from mechanics and thermodynamics to mechanothermodynamics* (Figs. 4.2 and 4.3). The fact that the both ways lead to one purpose and, finally, yield the same (unified) result, means that the above-mentioned two methodologies of analysis are valid, correct, and do not contradict each other.

It should be said that mechanothermodynamics research is just at the beginning. Deepening and expanding research in this new and perspective area of scientific knowledge is expected in the nearest future. The authors think that the usefulness of research is great and it is difficult to foresee this at present time.

Returning to Fig. 2.1, it is seen that it is ended with the arrow with a question: what sort of object will be behind the MTD system? The obvious and common answer is: it is our real world. Nowadays it is being studied actively by numerous and various sciences—from chemistry and biology … through mechanics and thermodynamics … and up to philosophy; from all points of view. From Fig. 2.1 it

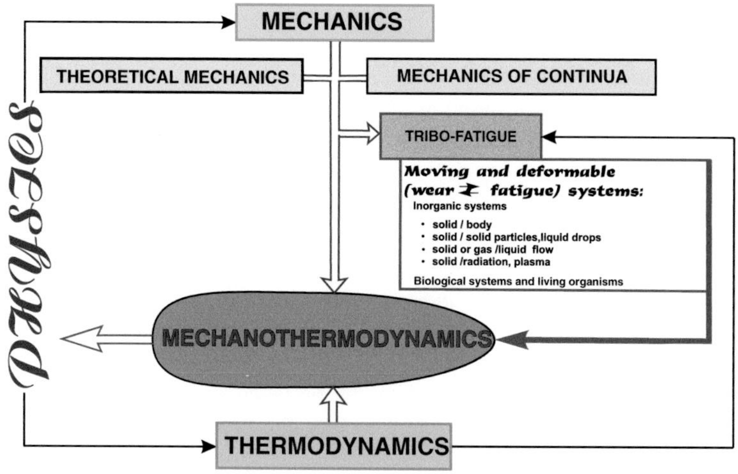

Fig. 4.2 Ways to mechanothermodynamics as a new branch of physics

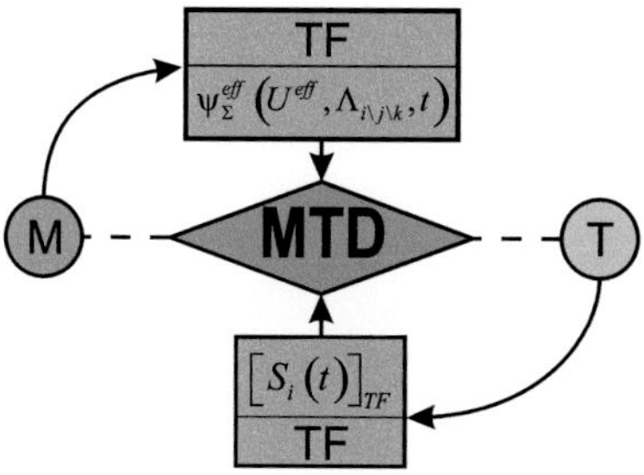

Fig. 4.3 Tribo-Fatigue bridges from mechanics (*M*) and thermodynamics (*T*) to mechanothermodynamics (*MTD*) (the *solid lines* with *arrows*; the *dashed lines* show the unrealized ways (for more than 150 years) from M or T to MTD)

is clear that the MTD system should be followed by an object that is somewhat more complex (however simpler) than a real system, for example, an MTD system with some "elements of intelligence." The first works in this area of research are already available [10, 24, 29].

4.2 First Principle of Mechanothermodynamics

All living things—animate and inanimate—*were born out of matter*, whereas *the motion is the mode of existence of matter* that will be represented as the objective reality given us in sensations and measurements; it is hence reflected in our mind. These philosophical postulates became the basis of current knowledge. One more fundamental postulate is formulated: *damage of everything that exists has no conceivable limits*. Having these three most general postulates, it is possible to talk about the *evolution of matter and, hence, about any systems to be formed by it* in the process of motion and corresponding interactions.

If the evolution is considered from the unified point of view of the mechanothermodynamic state of the system, then having regard of the above postulates, it is possible to understand: *not the thermodynamic death threatens any system, but the damage and decomposition into components that can and must in turn be considered as initial elements for the generation and development of new systems whose mode of existence is the motion*—new damages. *Evolution* appears to be hence *unidirectional in time* and *essentially infinite*, since matter as a mode of its existence—motion is *indestructible*.

Thus, the above-formulated postulate can be taken as the *first law of mechanothermodynamics* (Fig. 4.4). According to this, the "number of system damages" can be arbitrarily large:

$$\vec{\omega}_\Sigma = \vec{\omega}_\Sigma\left(U_\Sigma^{eff}\right) \overset{t}{\Rightarrow} \infty. \tag{4.1}$$

According to (4.1), the *evolution of any system inevitably involves the process of its decomposition,* in particular into a countless number of infinitely small components (atoms, elementary particles, etc.) (as already said, these components will become building bricks for new systems).

First principle of mechanothermodynamics

DAMAGE OF EVERYTHING THAT EXISTS HAS NO CONCEIVABLE LIMITS

$$\vec{\omega}_\Sigma = \vec{\omega}_\Sigma\left(U_\Sigma^{eff}\right) \overset{t}{\Rightarrow} \infty$$

$$d_\omega^* = e^{-\omega_\Sigma^*} \overset{t}{\Rightarrow} 0$$

The first principle states that the *evolution of any system is characterized by the inevitable unidirectional process of its damage and time decomposition* into an infinitely large number of small components (fragments, atoms, elementary particles, etc.). In essence, it is recognized that the evolution is endless, if decomposition products of any system are considered to become building material for new systems. In other words, our *Universe is indestructible since it evolves by damageability*. This corresponds to the philosophical concept that matter and motion are eternal, but *the damageability is the fundamental property and the duty of all systems*, including living and intelligent.

Conclusion **The production of internal mechanothermodynamic entropy is also eternal as motion and damage; this means that the entropy of the Universe grows.**

Entropy increase law

$$dS_{total} = \Lambda_{T\backslash TF}\left[(dS)_T + (d_iS)_{TF}\right] = \Lambda_{T\backslash TF}\left[\frac{dU + pdV}{T} - \frac{1}{T}\sum_1^n \mu dN_k + \gamma_1^{(w)}\frac{L_{\omega_\Sigma}}{T_\Sigma}dV_{P\gamma}\right]\uparrow$$

Fig. 4.4 First principle of mechanothermodynamics

As mentioned above, the fate of the system (or its longevity) is defined by the intensity and direction of interaction processes of irreversible internal damages caused by any actions. Thus, it is possible to formulate the *consequence of the first law of thermodynamics: internal mechanothermodynamic entropy production is eternal as motion and damage.*

The first principle of mechanothermodynamics can be essentially interpreted as the statement: the *Universe's entropy is growing*. At present, it is the accepted *philosophical law* of increasing entropy But if in philosophy this law is "claimed," then in mechanothermodynamics it is formulated in the form

$$dS_{total} = \Lambda_{T\backslash TF}\left[(dS)_T + (d_iS)_{TF}\right] = \\ = \Lambda_{T\backslash TF}\left[\frac{dU + pdV}{T} - \frac{1}{T}\sum_1^n \mu dN_k + \gamma_1^{(w)}\frac{L_{\omega\Sigma}}{T_\Sigma}dV_{P\gamma}\right] \uparrow .$$

In his known lectures on physics [3], *Feynmann* formulated the second law of thermodynamics in a similar manner. He started from the following consideration: irreversible thermodynamic changes are always peculiar for the system of Universe type. The same consideration was supported by *Schrödinger* [7], when he was thinking about what life meant. Remind that already in the nineteenth century, having introduced the concept of entropy, [2], *Clasius* came to the conclusion that became classical in modern thermodynamics [3]: Universe's (thermodynamic) entropy tends to a maximum. Since that time nobody could establish, at least, the order of the value of this maximum, but the statement itself was based only on a particular idea of thermodynamic entropy. But *Clasius' statement followed from the universal energy conservation law*: the Universe's energy is constant.

The statement that the Universe's energy is constant was based on the assumption that the Universe is an isolated system. If it is accepted that the Universe is an open mechanothermodynamic system that is able to exchange energy and matter with other cosmic supersystems, then the first law of thermodynamics acquires the physical reality.

4.3 Second Principle of Mechanothermodynamics

According to the *philosophical ideas, the interaction is a process of mutual influence of solids on each other*, it is a universal form of motion, development. The interaction specifies the existence and structural organization of any material system. The manifestation of all subjects and phenomena is the most general regularity of the existence of the world. The universal connection between phenomena has diverse manifestations. It includes connections between particular properties of solids or individual phenomena of nature that are expressed in specific laws, as well as such connections between universal properties of matter and development trends that manifest themselves in universal dialectic laws of life. *Any law is therefore a particular expression of the universal connection between phenomena*. The particular case of such interaction is the feedback in all self-regulating systems.

According to (particular) *physical concepts,* the interaction is the action of bodies or particles on each other that causes their motion to change. In *Newton's mechanics, the mutual action of solids on each other* is quantitatively expressed in terms of force. So, the *universal law of gravitation*

$$F = G\frac{m_A m_B}{r^2} \tag{4.2}$$

defines a force F of interaction between solids A and B with mass m_A and m_B, respectively, depending on the square of a distance r between their center; G is the gravitation constant.

It was proved that in the case of *electrically charged particles, the interaction is carried out through a mediator—electromagnetic field*. According to the *principle of reciprocity,* if a metallic solid with a constant electric charge Q_1 creates a potential φ_{12} on the second insulated metallic solid and the second solid with a charge Q_2 creates a potential φ_{21} on the first non-charged solid, then the principle of reciprocity is expressed as:

$$\phi_{21} = \phi_{12} \cdot \frac{Q_2}{Q_1}, \quad \text{or} \quad \frac{Q_2}{Q_1} = \frac{\phi_{21}}{\phi_{12}}, \quad \text{or } Q_1\phi_{21} = Q_2\phi_{12}. \tag{4.3}$$

Hence, the charge ratio is proportional to the charge-created potential ratio. *Generalizing the aforesaid, it is considered that the interaction between solids is carried out by these or those fields* (for example, gravitation—by a gravitational field) *continuously distributed in space*.

Tribo-Fatigue [15–17, 27] studies the *interaction between irreversible damages in the moving and deformable system*; the regularities of such interaction are due to a *field of stresses (strains) caused in it by appropriate Newtonian* internal force factors. Let, for example, σ, p, T be such factors that generate appropriate fields in the MTS system: thermodynamic (T) and strain (stress σ_{ij}, pressure p). The *law (Sosnovskiy's generalized rule) of interaction of damages* $\omega_\sigma, \omega_p, \omega_T$ *due to fatigue, friction and wear phenomena, changes in thermodynamic states* is then given in the form

analysis

$$F_\Lambda\left[\left(\omega_p \rightleftarrows \omega_\sigma\right) \rightleftarrows \omega_T\right] = \left[\left(\omega_p + \omega_\sigma\right)\Lambda_{\sigma\backslash p} + \omega_T\right]\Lambda_{M\backslash T} = \omega_\Sigma, \quad \Lambda \gtreqless 1 \tag{4.4}$$

synthesis

where the Λ's are the interaction parameters (functions) that can acquire three classes of values: $\Lambda > 1, \Lambda = 1, \Lambda < 1$.

According to (4.4), the analysis of the system (the integral ω_Σ thought as multiple $\omega_\sigma, \omega_p, \omega_T$) and its synthesis (many thought as the integer by the interaction functions of Λ) are possible. In this case, the analysis and the synthesis have a particular quantitative expression. This is the peculiarity and important advantage of law (4.4). Its second important peculiarity is the following. Here we are talking not about external forces, as, for example, in *Newton's* laws, but about irreversible damages of the system due to fields of internal forces. And, at last, the main thing is: *mechanothermodynamis must study and investigate not the mutual influence of factors but the interaction of phenomena*. Thereby, mechanics and thermodynamics are marked by a transition to the phenomenon analysis—in addition to the traditional factor analysis.

Bearing in mind principle (4.4), it is possible to understand and describe quantitatively (or assess comparatively) many phenomena known well (qualitatively) and also some new phenomena; the latter are called the surprises of Tribo-Fatigue [14, 21]. Here, let us give several examples, considering that any process $\omega_\Sigma(t)$ of damage accumulation in time t inevitably causes the system go to the limiting state (for example, to technical system failure) that is characterized by the critical damage ω_c (Table 2.4). This means that the limiting state criterion is of the form

$$\omega_\Sigma(t) = \omega_c. \tag{4.5}$$

Consider the situations when the values of ω_σ, ω_τ, ω_T are very small so that their simple sum is much smaller than the critical quantity ω_c. However, according to (4.4), it can be reached at $\Lambda \gg\ > 1$ (*strong interaction* of internal damages). So the *Tribo-Fatigue bomb* [14, 21] is realized—for example, anomalously low resistance to crack at fretting fatigue due a strong interaction of a set of weak damages. This is the *result of spontaneous softening of the system.*

Similarly, there also appears the *crowd effect* when "*individually weak*" people uniting for a specific purpose at appropriate interaction create critical situations in the society.

Further, let us briefly consider the situation when the quantities ω_σ, ω_τ, ω_T are large enough. So their simple sum is close, but slightly does not reach the critical value ω_C; in this case, $\omega_c - \omega_\Sigma = \Delta\omega \to 0$. There appears the *butterfly effect* when the slightest growth of the interaction function $\Lambda > 1$ can lead to severe consequences for the system; note that this effect can apparently be assessed quantitatively for the first time. Similar consequences can be generated *by* the *spark effect;* in this case, the limiting state is reached due to a small change in only one of the quantities: ω_σ either ω_τ, or ω_T.

Consider another situation: let the values of ω_σ, ω_τ, ω_T be so large that their simple sum exceeds the critical value ω_c. In this situation, the interaction function $\Lambda \ll\ < 1$; then the limiting state cannot be realized. It is the *collective effect* (or the *process of spontaneous hardening* of the system).

Note that in synergetic the *cooperative effect* is known, whose individual manifestations can be identified with those described above. But according to available information, the cooperative effect is not described quantitatively. Owing to this, this effect enables situations to be analyzed only qualitatively.

Principle (4.4) allows the so-called *transmitting states of the system* (see Sect. 4.2) to be predicted and described quantitatively. As a rule, these are the states of diverse and multiple failures due an unstrained growth of all components in

(4.4); these are the catastrophies and cataclysms of man-made and natural nature. In such cases, $\omega_{\Sigma}(t) \gg > \omega_C$.

And, at last, the main is: based on (4.4), analysis is made of *direct* and *back Tribo-Fatigue effects,* i.e., of the mutual influence of one (any) of the damages on changes in the remaining (two) damages.

Show that in such a case, the physical principle of reciprocity formalized, for example, in (4.2) or (4.3) is not satisfied. Indeed, according to (4.2), the interactions of bodies m_A and m_B are the same (at r = const) in the sense that the both bodies act upon each other with the same force $F_A = F_B = F$. Or, according to (4.3), two electrically charged particles interact in a similar manner. When irreversible damages interact, according to (4.4), the picture is fundamentally different. For example, at mechano-sliding fatigue (Tribo-Fatigue shaft /sliding bearing system), the friction and wear processes influence the fatigue resistance of the shaft (*direct effect*)

$$\sigma_{-1\tau} = \sigma_{-1}\sqrt{\frac{1}{\Lambda_{n\backslash\tau}} - \left(\frac{\tau_W}{\tau_f}\right)^2}, \tag{4.6}$$

whereas its wear resistance is formed under the influence of fatigue damages (*back effect*)

$$\tau_{f\sigma} = \tau_f\sqrt{\frac{1}{\Lambda_{\tau\backslash n}} - \left(\frac{\sigma}{\sigma_{-1}}\right)^2}, \tag{4.7}$$

so that the direct and *back* effects are quantitatively different, although the formulas for their calculation appears to be formally the same. The analysis showed that always, in all conditions of interaction for example between a shaft with a sliding bearing in terms of the field of strains (stresses) the inequality will be valid

$$\sigma_{-1\tau} \neq \tau_{f\sigma}.$$

The ratio $\sigma_{-1\tau}/\tau_{f\sigma}$ (or the ratio $\tau_{f\sigma}/\sigma_{-1\tau}$) turns out to be a *complex function not only of testing conditions and material properties* (σ_{-1}, τ_f), but also of the *conditions of irreversible interactions of damages* described by the appropriate Λ-functions with the account of $\Lambda_{n\backslash\tau} \neq \Lambda_{\tau\backslash n}$.

Despite the undoubted and diverse advantages, approach (4.4) has a substantial drawback: it is clearly formulated and written for three situations, phenomena, events $(\omega_{\sigma}, \omega_{\tau}, \omega_T)$. If their number is more than three, the similar solution (in terms of damageability) is not proposed according to the available information. In such a case, energy approach [27] appears to be effective. When this approach is used, the damageability is assessed more fully using the concept of an analyzed system as a mechanothermodynamic one, for which the effective (i.e., directly spent for damage generations) energy u_{Σ}^{eff}, associated with the entropy s_i.is determined.

As shown above and presented in [14–17, 21, 27], *flows of effective energy (entropy), caused by different-nature sources, at irreversible changes in the mechanothermodynamic system are not sumed up—they interact in a complex manner.*

Such Λ-interactions are described by the expressions

$$u_{\Sigma}^{eff} = u_{\Sigma}^{eff}\left(\Lambda_1, \ldots, \Lambda_m, u_1^{eff}, \ldots, u_n^{eff}\right), \quad m < n; \tag{4.8}$$

$$s_i = s_i\left(\Lambda_1, \ldots, \Lambda_m, s_i^{(1)}, \ldots, s_i^{(n)}\right), \quad m < n. \tag{4.9}$$

The result of multiple Λ-interactions is the development (accumulation) of internal damages in system elements that are defined by the unity and struggle of opposite physical hardening–softening processes. The interaction functions must therefore take three classes of values ($\Lambda \gtreqless 1$). Thus, the second law of thermodynamics can be formulated (Fig. 4.5).

Of course, of importance is the problem of creating the *general model of Λ-interactions*. To a first approximation, it can be built using *elements of graph theory*.

Second principle of mechanothermodynamics

EFFECTIVE ENERGY (ENTROPY) FLOWS CAUSED BY DIFFERENT-NATURE SOURCES ARE NON-ADDITIVE – THEY INTERACT DIALECTICALLY IN TIME:

$$U_{\Sigma}^{eff} = U_{\Sigma}^{eff}\left(\Lambda_1, \ldots, \Lambda_m, U_1^{eff}, \ldots, U_n^{eff}, \vec{t}\right), \; m < n$$

$$S_i = S_i\left(\Lambda_1, \ldots, \Lambda_m, S_i^{(1)}, \ldots, S_i^{(n)}, \vec{t}\right), \; m < n$$

The second principle states the driving force and the main source of the emergence and development of processes of internal damageability of any system – these are dialectic Λ-interactions of effective energy components in the absorbing medium.

Λ-interaction functions must take three classes of values ($\Lambda \gtrless 1$) in order to reflect not only *unity and struggle*, but also *direction* of *physical hardening-softening* processes in the system. Since hardening is always finite, whereas the softening intensity can be infinitely high, the interaction of such processes inevitably leads a system to *the critical* or *damageability-limited* state.

Conclusion **Effective energy absorbed in the system when it reaches a critical (limiting) state is identically equal to released (and dissipated) energy during its degradation up to decomposition.**

Effective energy conservation law

$$\int_0^{T_{\oplus}} u_{\Sigma}^{eff}(t)dt \equiv \int_{T_{\oplus}}^{T_*} u_{eff}^{\Sigma}(t)dt$$

Fig. 4.5 Second principle of mechanothermodynamics

The simplest expression for definition of effective energy (4.4) is of the form

$$u_{\Sigma}^{eff} = \left[\left(u_n^{eff} + u_{\tau}^{eff}\right)\Lambda_{n\backslash\tau} + u_T^{eff}\right]\Lambda_{T\backslash M}. \tag{4.10}$$

Structure (4.10) is substantially hierarchic: first, effective energy is determined at the interaction of its force and friction components

$$u_{n\backslash\tau}^{eff} = u_{n\backslash\tau}^{eff}\left(\Lambda_{n\backslash\tau},\ u_n^{eff},\ u_{\tau}^{eff}\right) = \left(u_n^{eff} + u_{\tau}^{eff}\right)\Lambda_{n\backslash\tau}, \tag{4.11}$$

then at the interaction of its mechanical and thermal components

$$u_{\Sigma}^{eff} = u_{\Sigma}^{eff}\left(\Lambda_{T\backslash M}, u_{n\backslash\tau}^{eff}, u_T^{eff}\right) = \left(u_{n\backslash\tau}^{eff} + u_T^{eff}\right)\Lambda_{T\backslash M}. \tag{4.12}$$

Write expression (4.10) in the general form:

$$u_{\Sigma}^{eff} = u_{\Sigma}^{eff}\left(\Lambda_{n\backslash\tau}, \Lambda_{T\backslash M}, u_n^{eff}, u_{\tau}^{eff}, u_T^{eff}\right). \tag{4.13}$$

Structure (4.13) can be presented in the form of a *graph tree* (Fig. 4.6a) or a *hypergraph* corresponding to it (Fig. 4.6b).

Expression (4.13) and the graphs shown in Fig. 4.6 are advantageous because they are valid for any energy operations, but not only for their summation as shown, for example, in (2.7).

The graph in Fig. 4.6 is a *rooted tree,* whose *leaves* are the "primary" effective energies $\left(u_n^{eff}, u_{\tau}^{eff}, u_T^{eff}\right)$, whose definition is rather arbitrarily, and the *vertices* are not leaves, but are the effective energies determined by the appropriate interaction functions $\left(\Lambda_{n\backslash\tau}, \Lambda_{T\backslash M}\right)$ and lower level energies.

The *tree root* is the *"total" effective energy* u_{Σ}^{eff}.

The *vertices* of the *hypergraph* shown in Fig. 4.6b *are the effective energies* $u_n^{eff}, u_{\tau}^{eff}, u_T^{eff}$, and its *ribs* are the interaction functions $\Lambda_{n\backslash\tau}, \Lambda_{T\backslash M}$. This hypergraph can be assigned by the following matrix:

	u_n^{eff}	u_{τ}^{eff}	u_T^{eff}
$\Lambda_{n\backslash\tau}$	1	1	0
$\Lambda_{T\backslash M}$	1	1	1

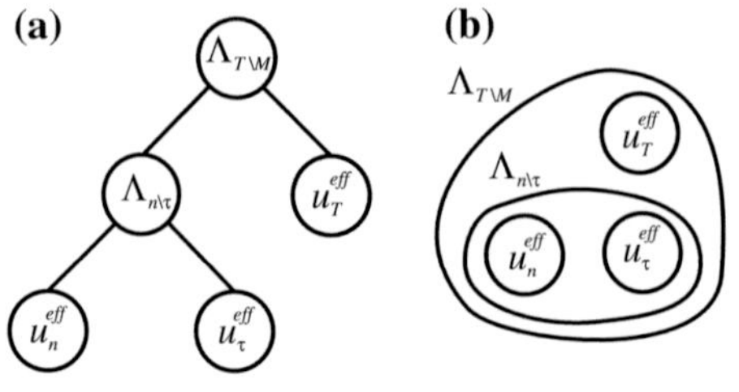

Fig. 4.6 Scheme of energy interaction

In the most general case, when primary energies are determined by some method, more than two (shown in Fig. 4.6) levels can be available; the functions Λ can determine the interaction of more than two energies. The effective energy for the system as a whole will be determined by

$$\left\{\Lambda_1, \ldots, \Lambda_q, u_1^{eff}, \ldots, u_s^{eff}\right\}, \quad q < s,$$

i.e.,

$$u_\Sigma^{eff} = u_\Sigma^{eff}\left(\Lambda_1, \ldots, \Lambda_q, u_1^{eff}, \ldots, u_s^{eff}\right), \quad q < s. \tag{4.14}$$

A detailed definition of expression (4.14) up to the second level can be given in the following form:

$$\begin{aligned} u_\Sigma^{eff} = \Lambda_\Sigma\Big\{\Lambda_1\Big[\Lambda_{11}(\cdots), \ldots, \Lambda_{1m}(\cdots), u_{11}^{eff}, \ldots, u_{1n}^{eff}\Big], \ldots, \\ \Lambda_k\Big[\Lambda_{k1}(\cdots), \ldots, \Lambda_{kp}(\cdots), u_{k1}^{eff}, \ldots, U_{kr}^{eff}\Big], u_1^{eff}, \ldots, u_l^{eff}\Big\}. \end{aligned} \tag{4.15}$$

Formula (4.15) in Fig. 4.7, *a* is presented in the form of the tree and its appropriate hypergraph that can be assigned by the matrix $R = \|r_{ij}\|$ where

$$r_{ij} = \begin{cases} 1, & \text{if } u_i \in \Lambda_j, \\ 0, & \text{if } u_i \notin \Lambda_j. \end{cases}$$

With regard to *Sherbakov–Sosnovskyi's model* as described above, it is apparently *possible to construct the general concept of* Λ-interactions *in any systems.*

The analysis of evolution of systems considering their damageability has one more advantage: there is an interrelation of motion and information in the system [27]. Moreover, to describe system states, other useful characteristics associated with damageability can also be adopted.

As an example, Fig. 4.8 (see also Table 2.4 and Fig. 2.21) illustrates the interrelations between *A*-, *B*-, *C*-, *D*-, and *E*-states of system damageability (Table 2.4) and typical unfavorable events (incidents, accidents, catastrophies and cataclysms) that can be described by changes in *risk and safety characteristics* [27]. Here, the *index L-risk* ρ is determined as the expectation of any unfavorable events, phenomena, situations A_1, A_2, …, A_n. If the probability of such events (situations) is

$$P(A_i) = P(A_1, A_2, \ldots, A_n) \equiv P(A), \tag{4.16}$$

and the probability Q of opposite—favorable events (situations) B_1, B_2,…, B_n is

$$Q(B_i) = Q(B_1, B_2, \ldots, B_n) \equiv Q(B), \tag{4.17}$$

then *Sosnovskiy's risk is determined* [18] *by a portion of "bad" in "good"*

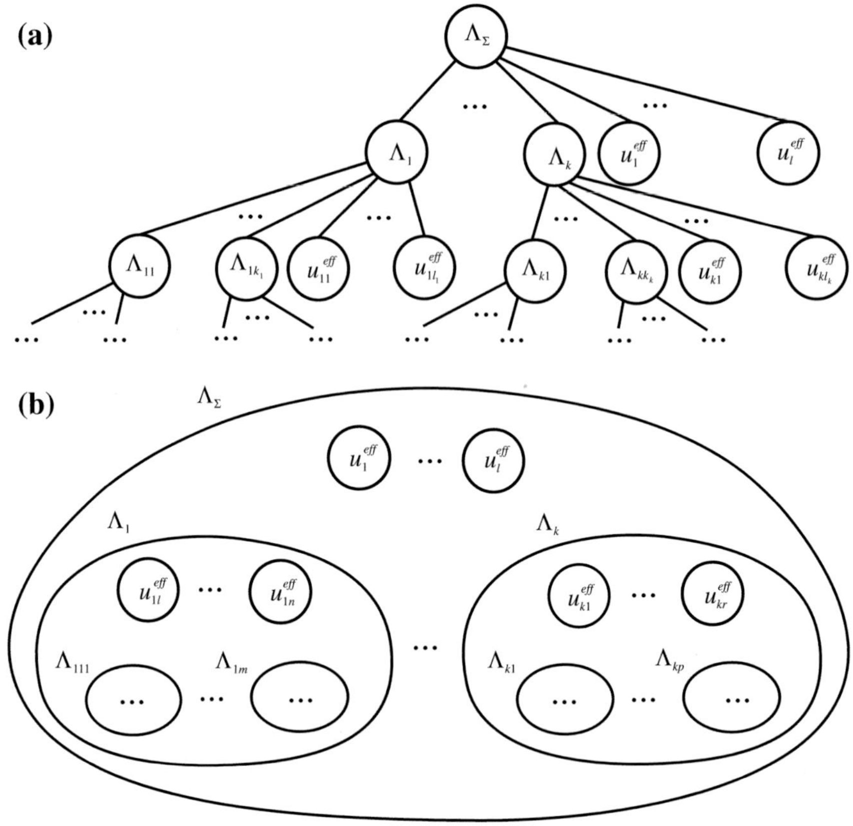

Fig. 4.7 General scheme of Λ-interactions

$$0 \le \rho(A_i, B_i) = \frac{P(A_i)}{Q(B_i)} \le \infty \tag{4.18}$$

provided that there is *some kind of the conservation law of opposites* in the form

$$P(A_i) + Q(B_i) = \text{const} = 1. \tag{4.19}$$

Considering that the *safety* S_ρ is the *opposite of the risk* ρ, we have

$$S_\rho + \frac{P(A_i)}{Q(B_i)} = 1. \tag{4.20}$$

Thus, formulas (4.18) and (4.20) describe the procedure of risk and safety analysis in the case of a set of events A_i that in fact cause it.

A more detailed analysis of the interrelation of risk and damage is given in [18].

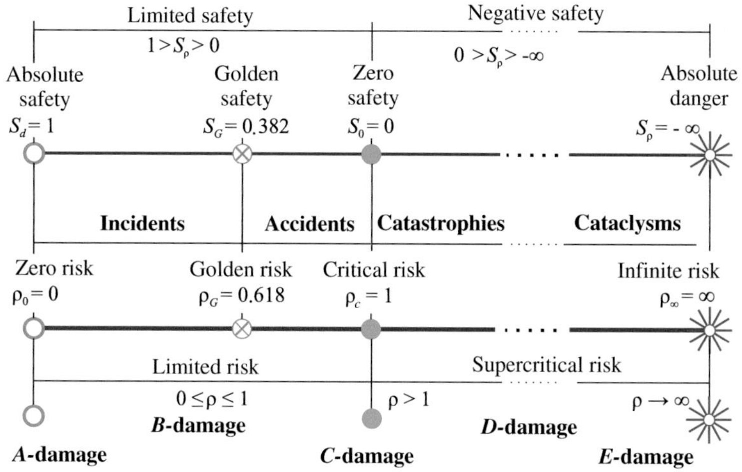

Fig. 4.8 Interrelation of system damages and the risk (safety of its existence)

4.4 Third Principle of Mechanothermodynamics

According to the first principle, the damageability of all things has no conceivable boundaries. Perhaps, this is the universal law of the Nature which in essence underlies the evolution of any systems—inorganic and organic, including living and intelligent. Indeed, if the dialectic damageability of solids is caused by hardening–softening processes, then similar but, of course, peculiar processes are revealed in liquid (for example, increase–decrease of viscosity), in gases (for example, increase–decrease of pressure), in organisms (for example, rejuvenescence-cytosis), etc. The second principle of mechanothermodynamics states the cause of damageability of systems—these are internal irreversible Λ-interactions of effective energy components due to different-nature sources. Further, it is natural that a question of damageability scales arises, because, as shown in Sect. 3.3, just dangerous volume sizes (Sect. 2.3) are responsible for the object state in terms of the generated entropy level. Hence, they directly characterize the damage of its functioning.

Let us explain this by the specific example. In Tribo-Fatigue systems, as a rule, not one but, at least, *several types of dangerous volumes* are revealed since different characteristics of the stress–strain state (1.1) (Table 4.1) are used for their definition.

Figure 4.9 is the graphical sketch of *component dangerous volumes* V_{xx}, V_{yy}, V_{zz}, their intersection (overlapped dangerous volumes) $V_{zz} \cap V_{yy}$, $V_{yy} \cap V_{zz}$, $V_{xx} \cap V_{zz}$, $V_{xx} \cap V_{yy} \cap V_{zz}$ and combination $V_{xx} \cap V_{yy} \cap V_{zz}$ (*tensor dangerous volume*). This figure also demonstrates the *tensor* $\psi_{ij}(dV)$ *of relative damaging stresses*. It is seen that for the equal value of the components of the tensor $\psi_{ij}(dV)$, the intersection of the three component volumes $V_{xx} \cap V_{yy} \cap V_{zz}$ is the most dangerous volume where the initiation of damage is expected. It is natural that in this zone, the primary

Table 4.1 Classification of static dangerous volumes

Dangerous volume type	Definition	Design formula	Relative damageability measure
Component	$V_{ij} = \{dV/\lvert\psi_{ij}\rvert \geq 1, dV \subset V_k\}$ The region of a loaded body, at each point of which the value of the corresponding component of the stress tensor is not less than the limiting value	$V_{ij} = \iiint_{\lvert\psi_{ij}(V)\rvert \geq 1} dV$	$\omega_{ij} = V_{ij}/V_k$
Main	$V_i = \{dV/\lvert\psi_i\rvert \geq 1, dV \subset V_k\}$ The region of a loaded body, at each point of which the value of the corresponding component of the principal stress tensor is not less than the limiting value	$V_i = \iiint_{\lvert\psi_i(V)\rvert \geq 1} dV$	$\omega_i = V_i/V_k$
Spherical	$V_S = \{dV/\lvert\psi_s\rvert \geq 1, dV \subset V_k\}$ The region of a loaded body, at each point of which the value of the component of the spherical part of the stress tensor is not less than the limiting value	$V_S = \iiint_{\lvert\psi_s(V)\rvert \geq 1} dV$	$\omega_S = V_S/V_k$
Deviator	$V_D = \left\{dV/\max_{i,j}\left\lvert\psi_{ij}^D\right\rvert \geq 1, dV \subset V_k\right\}$ The region of a loaded body, at each point of which the value of at least one component of the deviator part of the stress tensor is not less than the limiting value	$V_D = \iiint_{\max_{i,j}\lvert\psi_{ij}^D(V)\rvert \geq 1 \geq 1} dV$	$\omega_D = V_D/V_k$
Combined	$V_C = \cap_{i=p,j=m}^{q,n} V_{ij}$, $i, j, p, q, m, n = x, y, z$, $V_C = \cap_{i=p}^{q} V_i$, $i, p, q = 1, 2, 3$ The intersection of two or more dangerous volumes	$V_C = \iiint_{\bigwedge_{i=p,j=m}^{q,n}\lvert\psi_{ij}(V)\rvert \geq 1} dV$ $V_C = \iiint_{\bigwedge_{i=p}^{q}\lvert\psi_{ij}(V)\rvert \geq 1} dV$	$\omega_C = V_C/V_k$
Octahedral	$V_{\text{int}} = \{dV/\psi_{\text{int}} \geq 1, dV \subset V_k\}$ The region of a loaded body, at each point of which the value of the stress intensity is not less than the limiting value	$V_{\text{int}} = \iiint_{\psi_{\text{int}}(V) \geq 1} dV$	$\omega_{\text{int}} = V_{\text{int}}/V_k$
Tensor	$V_T = \cup_{i=p,j=m}^{q,n} V_{ij}$, $i, j, p, q, m, n = x, y, z$, $V_T = \bigcup_{i=p}^{q} V_i$, $i, p, q = 1, 2, 3$ The combination of two or more dangerous volumes	$V_T = \iiint_{\vee_{i=p,j=m}^{q,n}\lvert\psi_{ij}(V)\rvert \geq 1} dV$ $V_T = \iiint_{\vee_{i=p}^{q}\lvert\psi_{ij}(V)\rvert \geq 1} dV$	$\omega_T = V_T/V_k$

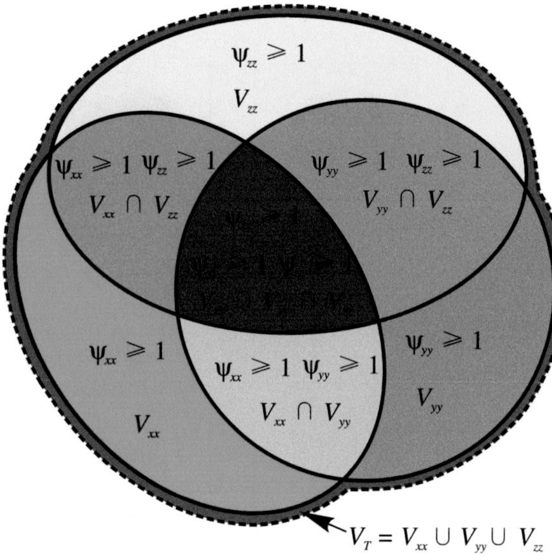

Fig. 4.9 Scheme of combination (*dashed line*)and intersection (*shaded*) of dangerous volumes caused by normal stresses

cracking most likely will start there where the tensor $\psi_{ij}(dV)$ is the largest in magnitude. Generally, the sketches like those shown in Fig. 4.9 allow diverse possible features of predictable damage to be analyzed. They are revealed, for example, in zones of intersection (overlapping) of dangerous volumes caused either only by normal or only by shear stresses or by a simultaneous action of normal and shear stresses of different sign. Thus, the possibility arises to specifically analyze the role of separation and shear processes in the formation of complex damage.

As an example, the calculation results for component dangerous volumes are presented in Figs. 4.10 and 4.11. From these figures it is seen that the dangerous volumes V_{zz}, V_{xz} and V_{yz} corresponding to the largest stresses $\sigma_{zz}^{(n)}$, $\sigma_{xz}^{(n)}$ and $\sigma_{yz}^{(n)}$ are the largest in size. Their further analysis permits the following two fundamental conclusions to be made. On the one hand, the *process of damageability* in the general case *is scattered,* i.e., the primary cracking can appear at any point of the dangerous volume. On the other hand, the *process of damageability appears to be discrete* in the sense that there exist local regions where the primary cracking is most probable (for example, in overlapped dangerous volumes—see Fig. 4.11, or in the multiply connected region of the tensor dangerous volume—see Fig. 4.10). Based on Figs. 4.10 and 4.11, it is also possible to make a deeper analysis of damageability as a consequence of the formation, development, and interaction of a set of dangerous volumes.

In engineering applications, to analyze the damageability caused by a complex (three-dimensional) stressed state at contact interaction, it is convenient to use the octahedral volume V_{int} (Fig. 4.12) as an integral characteristic. Comparing

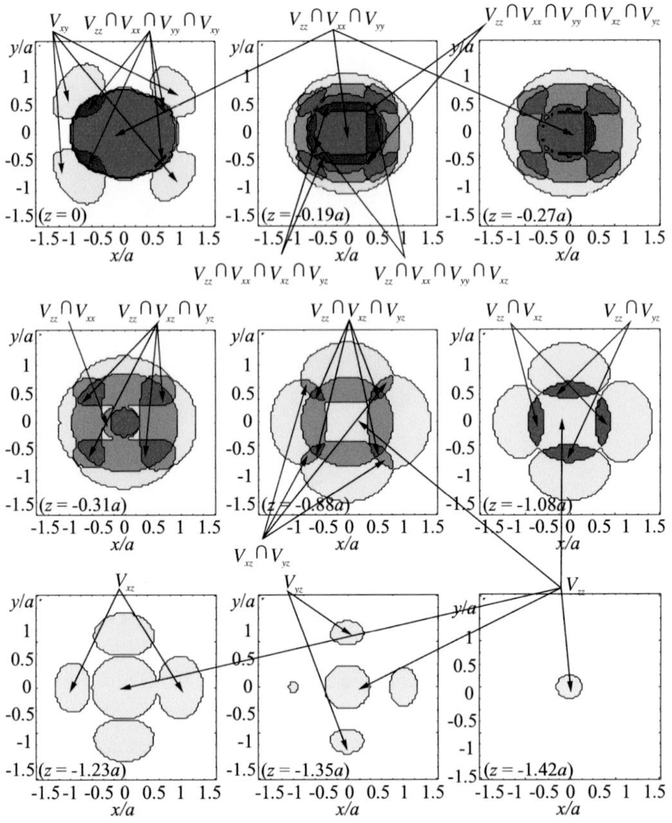

Fig. 4.10 Sections of the dangerous volumes V_{ij} in the rolling conditions under the action of $p(x, y)$, $q^{(||a)}(x, y)$ along the z axis for $\sigma_n^{(*\lim)} = 0.3p_0$, $\sigma_\tau^{(*\lim)} = 0.09p_0$, $f = 0.05$, $b/a = 0.813$

Figs. 4.12 and 4.13, it is easy to see differences in forms and size of dangerous volumes V_{int} for the cases of static and movable contacts. Moreover, these figures demonstrate a significant influence of the friction force $fp(x,y)$ on the changes in the dangerous volume form and size.

Using Figs. 4.12 and 4.13, *analyze the distinctive feature* of the distribution of damageability ψ_{int}. The distinctive feature of the ψ_{int} distribution is that a maximum value of ψ_{int} is on the z-axis under the contact surface. The value $\max\left(\psi_{\text{int}}^{(n)}\right)$ at frictionless contact is about 4.5. In the case of the most significant influence of the friction force on damageability, when it is codirected with the major semi-axis of the contact ellipse, a maximum value of ψ_{int} increases by about 60 %, traveling to the contact surface.

According to Figs. 3.1–3.4, a similar analysis can be made using the concept of energy dangerous volume.

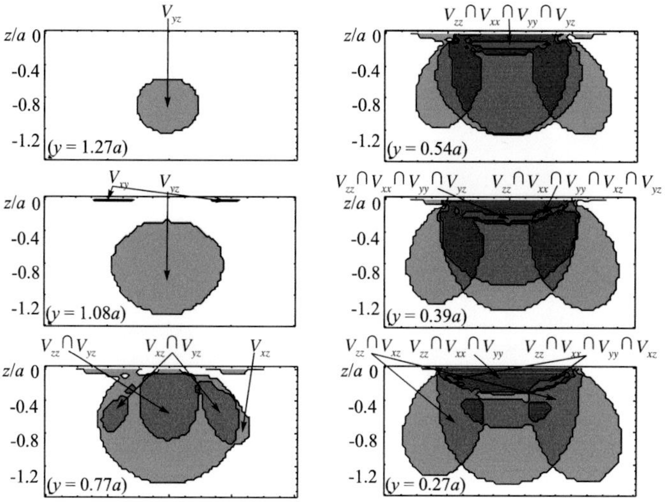

Fig. 4.11 Sections of the dangerous volumes V_{ij} in the rolling conditions under the action of $p(x, y)$, $q^{(||a)}(x, y)$ along the y axis for $\sigma_n^{(*\lim)} = 0.3p_0$, $\sigma_\tau^{(*\lim)} = 0.09p_0$, $f = 0.05$, $b/a = 0.813$

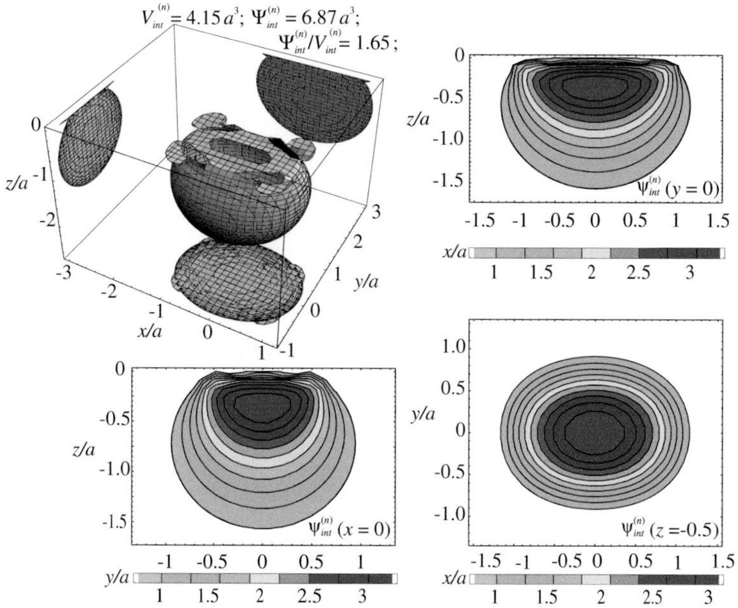

Fig. 4.12 Octahedral dangerous volume and damageability distribution in its different sections in the static contact conditions under the action of $p(x, y)$ for $b/a = 0.5$

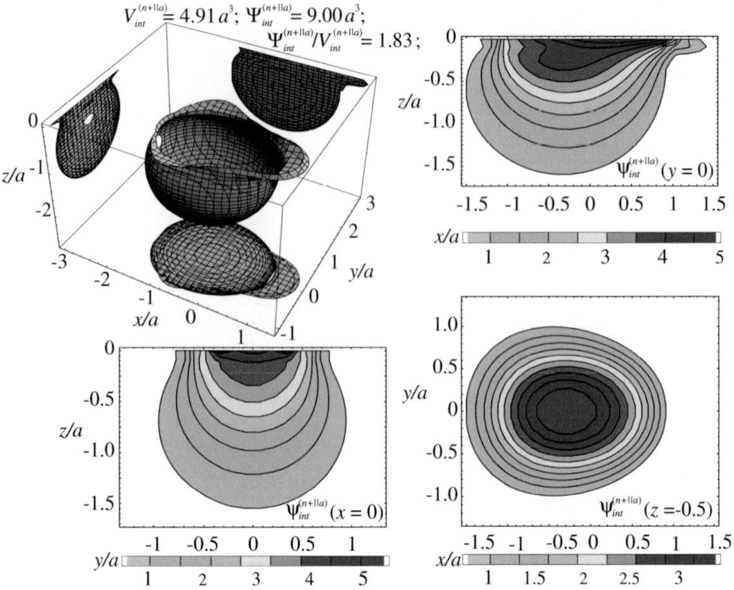

Fig. 4.13 Octahedral dangerous volume and damageability distribution in its different sections in the rolling conditions under the action of $p(x, y)$ and $q^{(\|a)}(x, y)$ for $f = 0.5$, $b/a = 0.5$

As the dangerous volume grows in an object during its functioning, the *critical (or limiting) stage* can be *reached*

$$V^{j}_{P\gamma} \in V^{j}_{crit}\left(U^{eff}_{crit}\left(S_i^{crit}\right), \vec{t}\right) < \infty, \quad j = 1, 2, 3, \ldots$$

From the aforesaid, it is clear that in the general case, we can talk about a *multiplicity of dangerous volumes* in MTD. Just this set of dangerous volumes characterizes the space of damages in it, i.e., their *scale*.

The generalizing description of the third principle of mechanothermodynamics is given on Fig. 4.14.

4.5 Fourth Principle of Mechanothermodynamics

As said above, if the first principle establishes a generalized principle of damageability of any systems, the second—its main cause, and the third—its scale, then the most important problem of finding the interrelation of motion, damage, and information arises. Its analysis (Sect. 3.4) just permits the fourth principle of mechanothermodynamics to be formulated. An extremely important question of the *emergence of intelligence*, for example, in specific biological systems (living and intelligent) requires further explanation.

Third principle of mechanothermodynamics

DEVELOPMENT OF IRREVERSIBLE DAMAGEABILITY PROCESSES IS POSSIBLE AND IMPLEMENTED WITH SOME PROBABILITY $P > 0$, WHEN IN A SYSTEM THERE DEVELOPS A FINITE REGION WITH A ZERO LEVEL OF EFFECTIVE ENERGY (INTERNAL ENTROPY) – DANGEROUS VOLUME

$$V_{P\gamma} \in V_{P\gamma}\left(U_{\Sigma}^{eff}(S_i), \vec{t}\right) > 0.$$

IF $V_{P\gamma} = 0$, THEN THE SYSTEM IS STABLE AND ITS EVOLUTION BY DAMAGEABILITY IS IMPOSSIBLE

The third principle allows *absolute value (size) of a damageability space* of a system to be established.
In the general case, a *set of dangerous volumes* is revealed in the system since their number is caused by many different-nature loads (mechanical, thermodynamic, electrochemical, etc.) and, hence, by many and various criteria for the transition to the dangerous state. Separate dangerous volumes can have *any sizes* (for example, nano-, micro-, meso-, macro-, giga-, etc.). Being overlapped, they interact dialectically when varying loads and/or time. The *scale of damageability* of the system in full characterizes the *damage (or risk) of its functioning*.
Reaching a critical (limiting) value of a dangerous volume and/or a damageability space gives rise to the *non-stationary behavior* of a system (for example, vibrations, whippings, etc.).

***Conclusion* The development and degradation of a system is the process of growing a space of multiple and multicriterial damageability in time**

$$0 < V_{P\gamma}^{j}(t)\uparrow < V_k, \quad j = 1, 2, 3, \ldots.$$

Fig. 4.14 Third principle of mechanothermodynamics

The thinking shows: since a living organism is composed of matter and a way of its existence is motion, the above results, in principle, can be applicable to it (of course, with necessary explanations) [19, 32]. According to the first principle of mechanothermodynamics, the damageability is a fundamental property and a required function of any systems—large and small, inorganic and organic, including living and intelligent. The use of this idea resulted in developing the concept (not considered here) of Tribo-Fatigue life of any biological objects (including man) [19, 21, 31, 32]. Briefly, it is formulated as follows: life is a special way of accumulation of irreversible damages. It appears that in any living and intelligent being, particular damages: biochemical, moral–psychological, intellectual, etc., occur and are accumulated. To date, the fundamental methods (not presented here) of learning such disorders—damages [14, 19, 32] are proposed.

Return, for example, to the function of dialectic transformation of the information of type (3.31), but as applied to some *dynamic system with elements of the intelligence* [18, 20]. From the simplest assumption it will be: *if the dynamic system has the start of the intelligence E_S, then its structure and behavior realize some control mechanism G*:

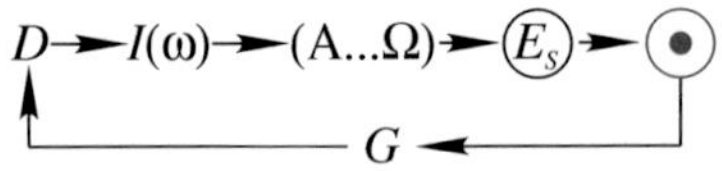

$$D \longrightarrow I(\omega) \longrightarrow (A...\Omega) \longrightarrow \widehat{E_s} \longrightarrow \bullet$$

$$\underset{\qquad\qquad\quad G \longleftarrow}{\rule{0pt}{0pt}} \tag{4.21}$$

Let us read (4.21) almost in terms of control theory. When *the dynamics D* of the system is changed, *new information I* (or its increment ΔI) is formed, for example, due to damageability processes $I(\omega)$; having properly transformed properly [for example, with the use of special alphabet $(A...\Omega)$, or with the use of the theory of image recognition, or ...], this information allows the *intelligence E_S*, which must realize a certain *goal* \bullet, to make a *necessary controlling action G*, whose intervention in the dynamics *D* of the system achieves the stated goal \bullet. *This is just the system of automatic control with the elements of the intelligence.*

It is easy to see that general task (3.30) can be slightly concretized with regard to (4.21):

$$_{\alpha}E\Omega \underset{I(\omega)}{\overset{Control}{\Leftrightarrow}} D \tag{4.22}$$

In (4.22) it is concretized that the intelligence elements and system dynamics are dialectically interrelated not only through information $I(\omega)$, but also through *G*; in what fallows, a direct understanding between the intelligence elements and system dynamics is not realized and is reached using a special alphabet-detector—α–Ω. It is assumed that this alphabet-detector from α to Ω must be an organic part of the intelligence E_S.

Where does the element of intelligence "come from"?

Information can be accumulated in time—so, for example, information function (3.30), (3.36) reads; its graphs are presented in Fig. 3.12. If such a function is smooth, then where does it direct? If it is not smooth, then what occurs with it in time?

Let us answer these questions from the *standpoints of dialectics*: we will consider that the universal law of transition from quantity to quality is also applicable to information. The process of information accumulation in time can then be presented as shown by the broken line in Fig. 4.15. Here, we will consider that *the information function is linear in the double logarithmic coordinates* ($\ln I - \ln t$), and its "smooth quantitative changes" are interrupted from time to time by *qualitative jumps* shown by the vertical line segments of type $1'1''$, $2'2''$, $3'3''$, etc. It is clear that the angle of slope α of the information function to the time axis must grow ($\alpha_1 < \alpha_3 < \alpha_2 \cdots$). Thus, if the critical information ΔI_2 is accumulated for a characteristic time, for example, measured by an interval Δt_2, then a *certain information function jump* δI_2 is realized; it just characterizes its transition into a *new quality* for the small time $\delta t_2^{(c)}$. Let us identify a new quality of the information function with the emergence of the corresponding elements of intelligence in the system. Thus, δI_2 is the *elementary conscious information*, whereas ΔI_2 is the *nonconscious information* accumulated for the time interval Δt_2.

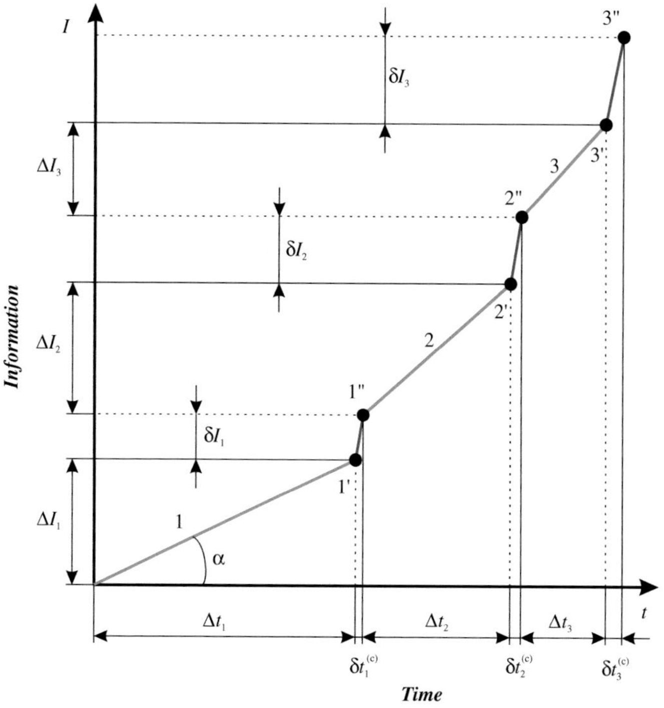

Fig. 4.15 Dialectic information accumulation function in time

If in the general case, the characteristic time period Δt_j $(j = 1, 2, \ldots, n)$ corresponds to the quantity of the accumulated (nonconscious) information ΔI_j, then some interval $\delta t_j^{(c)}$ must correspond to the information jump δI_j.

Hence, for any characteristic time period (j = const), the relations are valid

$$\frac{\delta I_j}{\delta t_j^{(c)}} = \frac{\Delta I_j}{\Delta t_j}, \tag{4.23}$$

or in other form

$$\delta I_j \Delta t_j = \Delta I_j \delta t_j^{(c)}. \tag{4.24}$$

According to (4.24), the product of the information jump (δI_j) by the time of its preparation $\left(\Delta t_j\right)$ is equal to the product of the accumulated (nonconscious) information $\left(\Delta t_j\right)$ by the time $\left(\delta t_j^{(c)}\right)$ of the information jump. In other words, according to (4.24), the increments of quantitative $\left(\Delta t_j\right)$ and qualitative (δI_j) information based on its characteristic time (Δt_j and ΔI_j, respectively) are equal

$(\text{tg}\,\alpha_j)$ and constant (at j = const). In essence, tg α characterizes the quantity of accumulative information per unit time, i.e., the logarithmic information density:

$$\frac{\Delta I}{\Delta t} = \Delta i. \tag{4.25}$$

Either from (4.23) or from (4.24) we have a formula for a *quantitative assessment of the information jump*

$$\delta I_j = \frac{\delta t_j^{(c)}}{\Delta t_j} \Delta I_j = \chi_j \Delta I_j, \tag{4.26}$$

where

$$\chi_j = \frac{\delta t_j^{(c)}}{\Delta t_j} = \frac{\delta I_j}{\Delta I_j} \tag{4.27}$$

is the *dimensionless parameter of information quality*. Its meaning is established as follows. If $\delta t_j^{(c)} = 0$, then $\chi_j = 0$ as well, i.e., the elements of intelligence are absent at $\chi_j = 0$. A very large numerical value of the parameter $\chi_j = 1$ is reached when the entire accumulated information is "converted" (dialectically transformed) into the intelligent one $(\delta I_j = \Delta I_j)$. Thus, the value of the information quality parameter is within the interval

$$0 \leq \chi_j \leq 1, \tag{4.28}$$

and its specific value $\chi_j \ll 1$ characterizes a share of the conscious (intelligent) information in the total quantity of the accumulated (nonconscious) information. Owing to this, parameter (4.27) is pertinently called the *index of intelligence*.

It should be noted that all parameters in formulas (4.23)–(4.27) are logarithmic and positive, i.e., for example,

$$\Delta t = \ln t_2 - \ln t_1, \quad t_2 > t_1,$$
$$\Delta I = \ln I_2 - \ln I_1, \quad I_2 > I_1,$$

since information functions are being studied at positive time when $0 \leq \alpha \leq 90°$. But if $270° \leq \alpha \leq 360°$, then information function will be *negative* (within positive time). Earlier, we have already obtained the negative information function from entirely different considerations (Fig. 3.12). This gives reason to think that it is real.

From Fig. 4.15 it is seen that the *increment of the elementary intelligent function* is not continuous in time and *occurs discretely* (from time to time) differing quantitatively at each stage (j = 1, 2, …) of development. But if this occurs in the same system, then it can be assumed that elementary quantities $(\delta I_j), j = 1, 2, 3, \ldots,$

Fourth principle of mechanothermodynamics

MOTION PROVIDES NEW INFORMATION IN A SYSTEM, IF THE INDEX OF ITS DAMAGEABILITY IS NONZERO (j 0); INFORMATION APPEARS TO BE EITHER POSITIVE, WHEN A SYSTEM IS HARDENED, OR NEGATIVE, WHEN IT IS SOFTENED. INTERRELATION OF MOTION, DAMAGE, AND INFORMATION IS DEFINED BY RECIPROCITY RELATIONS

$$\Delta I(t) = -k \cdot trB \cdot t = a_S \Sigma \omega_j(t).$$

The fourth principle states that *information sources in the Nature are matter and its damages in motion*. External forces and internal -interactions are the cause of motion and damage of material bodies and any systems, composed of them. Their long-term action in time causes the corresponding *information accumulation*. A *jump-like growth of information* is possible at singular (critical) points of development, when the quantity of accumulated information becomes critical. Any jump of information means the change in *its quality*.

***Conclusion* During motion and damage of a living system the jumps of information ($\pm\delta I$) generate the elements of intelligence (E_s) in it. Their algorithm quantitative assessment is possible.**

$$\dot{X}\left(\text{div}F(\bullet), U_{\Sigma}^{eff}, \omega_j, t\right) \begin{matrix} \nearrow^{+\delta I} \searrow \\ \searrow_{-\delta I} \nearrow \end{matrix} E_S$$

Fig. 4.16 Fourth principle of mechanothermodynamics

must be accumulated and combined somehow, apparently, interacting in a complex manner. The study of the problem of such interactions is beyond the scope of the present work; the book [23] contains additional considerations on this subject.

The generalizing description of the fourth MTD principle is given on Fig. 4.16.

4.6 A-Evolution

According to the available information, the general theory of evolution of systems has not been developed up to now. Indeed, in mechanics an attempt to develop such a theory was not a success, since equations of motion of (mechanical state) systems do not distinguish the past ($-t$) and the future ($+t$); in thermodynamics the theory predicts the heat death of the Universe, which is not consistent with modern scientific views; in synergetics studies are mainly concerned with the self-organization of systems in different nonequilibrium states; in dialectics only the most general qualitative behavior of systems is formulated.

We make a new attempt to construct the *fundamentals of the general theory of evolution* on the basis of the fundamental concepts of *damageability of systems* developed in Tribo-Fatigue [8] and the original *concept of L-risk* [18] organically bound with *golden proportions* [26]. Here, we can talk about any systems—inorganic, organic, living, and intelligent.

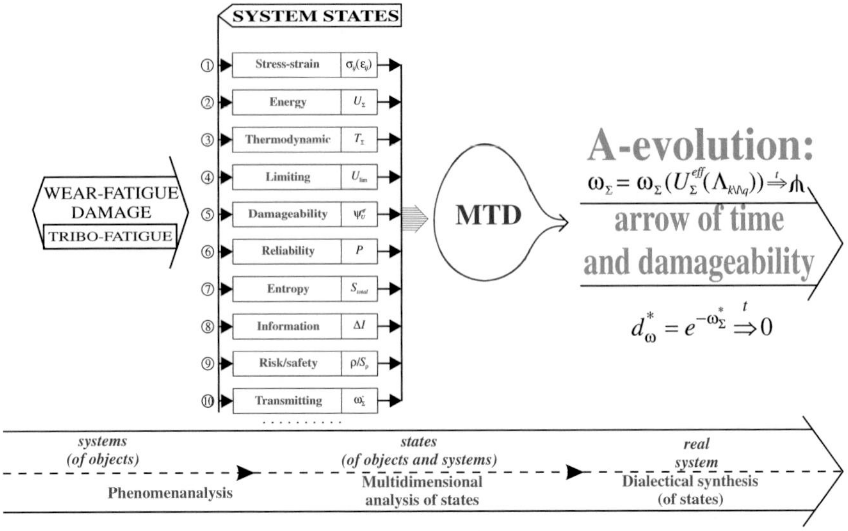

Fig. 4.17 Hierarchic structure of learning the macrocosm of matter: from simple to complex

The development of such a theory required, first of all, to work out the research methodology of systems [20]: from factor analysis through phenomenon analysis to dialectic synthesis (Fig. 4.17). As known, in the second half of the twentieth century a transition from the research of individual objects (considered material world) to the study of their systems [1] was proposed and realized. Then there came the inevitable stage of further integration: as a result of *interdisciplinary studies* (the principle of which is described, for example, in [5, 11]), a *new and perspective branch of mechanics—Tribo-Fatigue* [15–17] was developed. It appeared that if first the *factor analysis* was the basic methodology, then later a *new methodology—phenomenon analysis* (Fig. 4.17) was required (and formulated in [19, 27]). Its essence is outlined elsewhere in [4, 9, 19, 23, 27, 34]. It is found that friction and wear (on the one hand) and fatigue (on the other) are not factors but phenomena, and these *phenomena interact dialectically* in the operation of *specific systems* that are called *active* (according to GOST 30638–99), or *Tribo-Fatigue ones*. Of course, the phenomenon analysis "does not abolish" the use of the factor analysis methods, but it provides an adequate description of the behavior of a complex Tribo-Fatigue system under test and operation conditions [22, 28]. We think of Tribo-Fatigue as "the whole conceivable as much in relation to sciences associated with it, including with respect to those that were its progenitors. Not a mutual influence of factors but the interaction of phenomena—it is what is being studied by Tribo-Fatigue" [27].

Dialectic synthesis: the representation of a set of events, states, processes (in nature and society) as the generalized whole with characteristic integral properties and functions caused by a set of phenomena.

It is natural that nowadays sciences continue to integrate and unite. The main feature of a further improvement of the research methodology is that *from the analysis of objects and systems, a transition has been made to the analysis of their states* (Fig. 4.17), since it was found that different objects and systems "experience" the same states. The *multivariate analysis of typical states* (Fig. 4.17) is the result of *transdisciplinary studies* (the principle of which is described, for example, in [6]). It is easy to see that: this is a *qualitatively new stage of development of sciences*. Hence, from here we "are very close" to developing a *new physical discipline* (Figs. 4.1 and 4.2)—*mechanothermodynamics* (MTD), which as described in Sect. 4.1, was born *at the junction* of mechanics and thermodynamics [12–14]. This happened when *two bridges* were built (Fig. 4.3 and Sect. 4.1). One bridge is *Tribo-Fatigue entropy* [30] that paved the path *from thermodynamics to mechanics*. Another bridge is the *fundamental concept of irreversible damageability of all which exists* [19, 22, 32]. It paved the way *from mechanics to thermodynamics*. This path and this way are mutually penetrated by dialectic Λ-*interactions* [4, 9, 23, 27, 33] between damages caused by different-nature loads (mechanical, thermodynamic, electrochemical, etc.) and by characteristic entropy components (thermodynamic, Tribo-Fatigue, chemical, etc.). *The mechanothermodynamic system as a typical and important component of the real world* and its being (Fig. 4.17) thus become objects of study in the natural science and requires a philosophical understanding [18]. So it was time to develop the *general theory of evolution* that was called the A-*evolution* [18]. It was developed as a result of the multivariate analysis of states through their dialectic synthesis (Fig. 4.17) with the use of the principle of *convergence: an irreversible change in any states is generally interpreted as internal damage*. Hence, the A-*evolution is the evolution of any system by damageability* (Table 2.4). It is *characterized by the common and unidirectional arrow of time and damageability* (Fig. 4.17). As a result of the A-*evolution* the system inevitably dies, or it decomposes into particles of small size ($d_\omega \rightarrow 0$), for example, into atoms, elementary particles, etc. In other words, the degree of its damageability becomes infinitely large ($\omega_\Sigma \rightarrow \infty$). Based on these preconditions, it appeared possible to write the *known philosophical law of entropy increase* in the *form of the simplest equation* (Fig. 4.4) using the concepts of the thermodynamic entropy S_T (energy dissipation characteristic) and the Tribo-Fatigue entropy S_{TF} (energy absorption characteristic). Figure 4.4 illustrates a *fundamental difference* of the laws of thermodynamics and mechanothermodynamics. If the first describes the tendency of the body system A_1–A_2 to *thermal equilibrium* (the horizontal dash-dotted lines in Fig. 3.19), then the second predicts the *endlessness of damageability processes*: it is naturally assumed that decomposition products of any system become a building material for new systems. So the *evolution hysteresis* is realized.

The fundamentals of the theory of evolution by damageability (Sect. 2.9) should then be clarified in the aspect of the *reproduction of systems* It seems that the *past* (-*t*) and the future (+*t*) of the system (expressions (2.83)) *are radically different* as the *present* differs from the past and the future. Is it true?

Using Fig. 4.18, let us trace the *fate of any system of matter*—either inorganic or organic, including living and intelligent—"*beginning to end,*" i.e., from the beginning of existence in the past, through the being in the present and to the end of

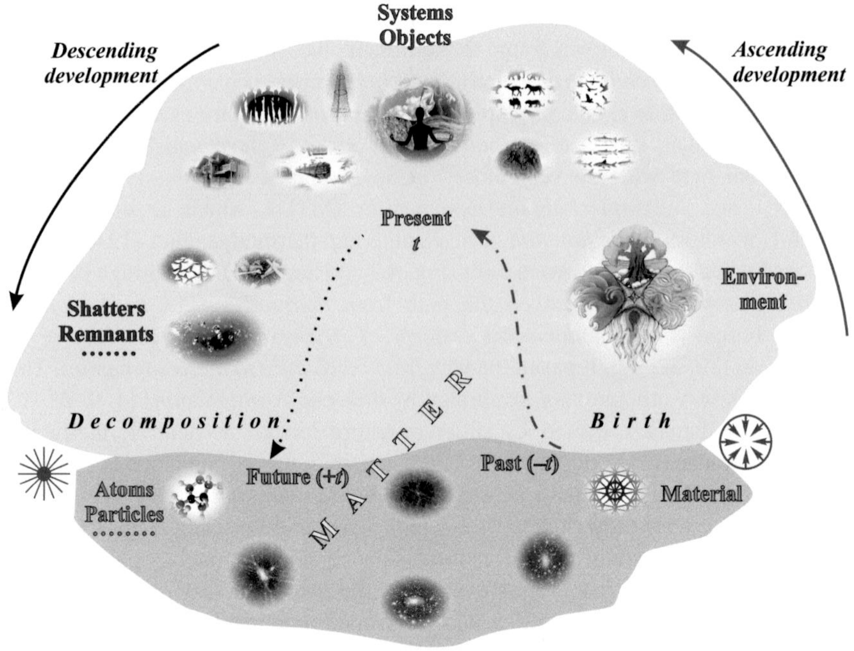

Fig. 4.18 Explanation of the evolution hysteresis: some conversions in the Earth/near-Earth space system

a subsequent (new) existence in the future. It is not difficult to understand that a *full cycle of evolution* of systems is determined by a succession and a set of their *characteristic states*. *Features of existence and being of specific systems* can be described as their determined states, for example: mechanical, thermodynamic, biochemical, intellectual, psychoemotional, public, information, entropy, energy, electromagnetic, risk and safety, mechanothermodynamic, damageability, etc.

Thus, the *state* is a *defined set of properties and functions* characteristic of a studied (perceived, measured) object of matter of given length and form at a given time moment. A change in the state is hence determined by the ratio and struggle of opposites responsible for the *existence and being of a given system* under particular conditions. But whatever are these conditions—irrespective of the features of existence and being of systems—their beginning and end are one and the same thing: a *matter* or, more precisely, a *matter state* (Fig. 4.18). Here, the concept of *matter* is defined as an *objective reality of any space length given to intelligent man in his sensations and measurements in time which are just being reflected in his consciousness*. As seen, our definition of matter goes back to its traditional understanding in philosophy, but it is somewhat wider and, perhaps, more accurate since it involves measuring the states of material bodies.

From the abovesaid, it is clear that the past and future of any system *do not* differ drastically; finally, the *past and future* of the system *are one and the same thing—it is matter*. This is just the essence of the *evolution hysteresis*. But from Fig. 4.18 it clearly follows that the matter, from which a given system was born, is not the matter, in which, at last, it has evolved, because *the world of matter is variable in time and space due to infinite motion*. Just as man cannot enter twice into the same river, so even one atom of a system cannot get out of the past and cannot be back to the future at the same place of the "*flow of matter*."

Examining the above short list of some possible characteristic states of any systems, once again we come to the conclusion: *only states of damageability are fundamental* since they accompany any system "from beginning to end," i.e., throughout its evolution at any time moment of its existence and being. All other states are only *temporal;* they are characteristic and important only within certain intervals of the life cycle of the system. That is why, it is important to give, possibly, a full definition of the concept of damageability. Earlier, it has been introduced in Item II (page 10). Now it can be generalized: *damage is an irreversible change of any state and, hence, a relevant function of the system as a whole or objects of its components*. It is important to emphasize once more that *any damage is real since it can be perceived, seen, touched, measured; it is objective as it exists and develops irrespective of the intelligent man's consciousness, but it is only reflected in it*. Thus, as said in the first MTD principle, *damageability is* a *fundamental property* and *a duty function* of both any matter system and its any component.

Based on the aforesaid, let's present the definitions of some of our concepts:

- *fundamental property of matter*: the ability to reproduce itself upon any transformations (change of any states);
- *fundamental property of space*: the ability to transform states of any extension and form; its any defined form is only temporal;
- *fundamental property of time*: the ability to create a continuous unidirectional flow; it can be stopped only mentally;
- *the present* is the time, space, and form of the being of the defined systems observed by intelligent man (perceived, measured, and then reflected in his consciousness)—current, particular systems (objects) of the world of matter;
- *the past is* the time, space, and form of existence of present (particular) matter systems (object) in other—previous states of the material world;
- the *future* is the time, space, and form of existence of present (particular) matter systems (object) in new—subsequent states of the material world.

It should be noted that we refer the concept of *being* only to the present and the concept of *existence*—only to the past and the future. Only moments of intermediate states of studied systems (between the past and the future) are perceived and reflected in our consciousness. In other words, the present is only an instant—a time limited by these or those considerations of intelligent man; it can be both "small" and "large" (long). Thus, our views of the present, the past, and the future of a system under study are subjective, although they reflect its objective existence during some time period.

In summary, we can say that *matter is the common past, present, and future of all that exists*. In our opinion, it is an *objective reality of the Universe*. This is just exhibited through the *fundamental essence of the evolution hysteresis of the material world*.

Figure 4.17 has no timescale. Our qualitative concept of the existence time of a system is shown in Fig. 2.23; now, we supplement it with the quantitative analysis (Fig. 4.19). Here, the past and the future are conditionally separated from the present by the shaded boundaries. It was important to make a quantitative description of a full interval of that time, during which any system exists, if we assume that its birth (emergence) corresponds to the time $t = 0$ and its death—to the time $t =$ ሐ (big) (Fig. 2.24). Indeed, the above proposed (Sect. 2.9) time interval [0; ሐ], in whose writing the square brackets mean its extreme values included into the consideration of the time interval itself. But it should be repeated that for other systems, this time interval will be the same; however, it will be located on other segment both of the world time (Fig. 4.19) and with respect to the existence time of any other system (Fig. 2.21). Thus, the *existence timescale of* a system shown in Fig. 4.19 appears to be *universal,* i.e., valid for any system.

An important feature of this scale is that here, the absolute time unit ($t = 1$) is set, which distinguishes into two fundamentally different domains of its numerical values: (a) $t < 1$ and (b) $t > 1$. It is considered here that for a given (particular) system neither "infinitely small," nor "infinitely large" time is realized since physical limits exist $t_0 = 0 = \lim\limits_{n\to\infty} 10^{-n}$, $t_{\text{ሐ}} = \text{ሐ} = \lim\limits_{n\to\infty} 10^{n}$

As seen, here the "eternal problem" is touched upon the concept of infinity, and its particular solution in a specific case is proposed. Of course, this solution is controversial.

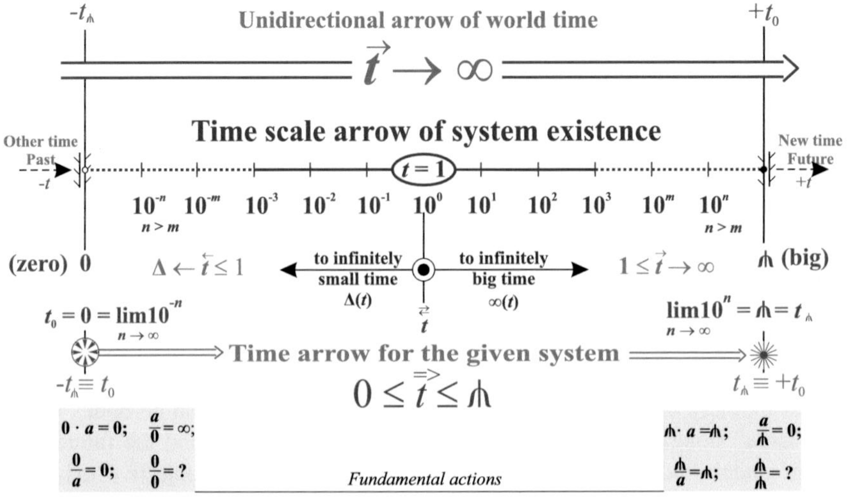

Fig. 4.19 Timescale hypothesis

It is of interest to note that the time of the domains $t < 1$ and $t > 1$ is substantially different and the corresponding scales are therefore antisymmetric.

The upper line in Fig. 4.20 illustrates the aforesaid schematically, generally, and graphically. The second line needs to be discussed: it is the question about the *evolution strategy*. Indeed, if the elements of intelligence occur in a system and if they are able to accumulate, then it means that some (*intelligent?*) strategy of system must occur and be implemented. Hence, it must have an *objective function*. The second line in Fig. 4.20 shows the analysis of this function [18, 23] based on the learning of *risk/safety states* of a system (see ⑨ in Fig. 4.17 and also Table 2.4 and Fig. 4.8).

Following [18, 23], let us consider two (I) and (II) evolution strategies.

The evolution involving the strategy of S_ρ-safety, i.e., is

$$\frac{S_\rho}{\rho}(t) = \text{Evolution (I)}. \tag{4.29}$$

In such a case, the *striving for absolute safety*

$$\frac{S_\rho}{\rho}(t) \Rightarrow S_d = 1. \tag{4.30}$$

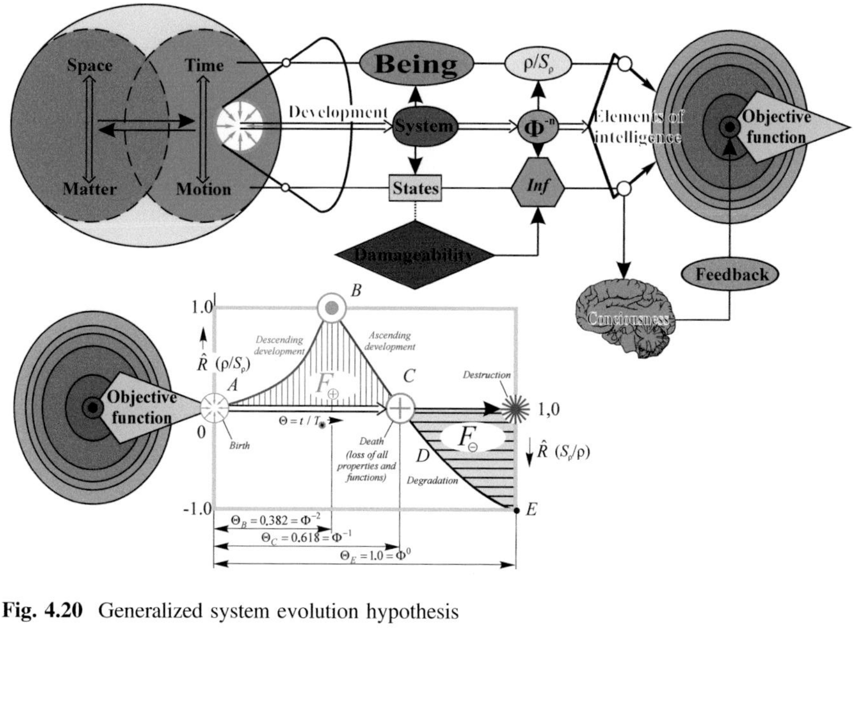

Fig. 4.20 Generalized system evolution hypothesis

is recognized as the universal. Herewith, the state of the system at the beginning of evolution ($t_0 = 0$) is described as follows:

$$\frac{S_d}{\rho_0} = \infty; \quad \rho_0 = 0; \quad S_d = 1; \quad \theta = \frac{t}{T_*} = 0,$$

where T_* is the absolute existence time of a material system (from birth to death).

The system *death* ($t = T_*$) is governed by the conditions

$$\frac{S_\infty^{(-)}}{\rho_\infty} = -1; \quad \rho = \infty; \quad S_\rho = -\infty; \quad \theta = 1.$$

Thus, the *relative range of* the existence time of the system is

$$0 \leq \frac{t}{T_*} = \theta \leq 1. \tag{4.31}$$

Similarly, the *evolution involving the strategy of L-risk* is

$$\frac{\rho}{S_\rho}(t) = \text{Evolution (II)}. \tag{4.32}$$

In this case, the *striving for the zero risk*

$$\rho(t) \to \rho_0 = 0$$

is recognized natural for the Nature.

In this case, the system state at the beginning of evolution is described by the following parameters

$$\frac{\rho_0}{S_d} = 0; \quad \rho_0 = 0; \quad S_d = 1; \quad \theta = 0,$$

and at the death of the system

$$\frac{\rho_\infty}{S_d^{(-)}} = -1; \quad \rho = \infty; \quad S_\rho = -\infty; \quad \theta = 1.$$

Work [18] proposes the power function for evolution (I) and, hence, the function inverse to it for evolution (II). Then having allowed for the above boundary conditions, the generalized function *ABCDE* graphically shown in Fig. 4.20 (second line) can be obtained. Wherein the functions of the physical state of the material system will be of the form:

$$\hat{R}(\theta) = \frac{\rho}{S_\rho}(\theta) \quad \text{at section } AB,$$

$$\hat{R}(\theta) = \frac{S_\rho}{\rho}(\theta) \quad \text{at section } BCDE.$$

$$\left.\begin{array}{r} 0 \le \theta \le 1; \\ -1 \le \hat{R}(\theta) \le +1. \end{array}\right\}$$

It is surprising, but it is the fact: the "*entire evolution*" *of any system "goes" into a single—fundamental rectangle with the boundary coordinates* $\pm 1, 0$. The discussion of this fact goes beyond the scope of the present work. The analysis showed [18] that the existence time of the system is described as three-stage (Table 4.2). As found, the relative time θ of each of the stages corresponds to *golden numbers* Φ (Phidias) [18].

Areas under the curve *ABCDE* are characterized by *energy functions of action*

$$\hat{F}(\theta) = \int \hat{R}(\theta) d\theta.$$

The death index of any system is

$$\psi_F = \left|\frac{F_+}{F_-}\right| = 1 \pm \Delta_c, \quad \Delta_c \to 0.$$

The action functions $F_+ = F_-$ can be equal only when the known laws of conservation of energy and matter are undoubtedly and rigorously satisfied: how much energy and matter have appeared (in some forms) to the death moment—so much energy and matter disappeared (in other forms) after the death of a given system. This corresponds to law (2.89) that turns also to be the consequence of the second principle of mechanothermodynamics (Fig. 4.5).

In summary, generalizing the above-mentioned and available experimental results, the following conclusion can be made. *If Nature as Man confesses the risk and safety strategy, then naturally they should have what is called intelligence*. The form and content of the Nature's intelligence is drastically different from the Man's intelligence—this means inescapable difficulties in its understanding by those who live on the Earth and have certain-type consciousness [18].

Table 4.2 Evolution stages

Stages	Physical states	Energy states	Damageability stages
Stage I	$0 < \hat{R}(\theta) \le 1$	$F_+ > 0$	$0 \le \omega_{\Sigma\theta} < 1$
Stage II			
Stage III	$-1 \le R < 0$	$F_- < 0$	$1 < \omega_{\Sigma\theta} \le \infty$

Much has been written about the intelligence. Maybe, the author of work [18] was the first who guessed for what it is needed: *to maintain a required balance between safety and risk in terms of the damageability criterion of any systems*: large and small, living and nonliving *so that their existence time would obey golden relations*. How much this hypothesis is true, future research will show.

All of us revolve around the truth. But the point is get closer to it and not away from it. The authors do hope that in this respect, some step is made in the discussion of this problem.

References

1. Blauberg, I.V.: Integrity problem and the systems approach. Moscow (1997) (in Russian)
2. Clausius, R.: Mechanical Theory of Heat. John van Voorst, London (1867)
3. Feynman, R.: Feynman's Lectures on Physics. Mir, Moscow (1963) (in Russian)
4. Frolov, K.V., Makhutov, N.A., Sosnovskiy, L.A.: Theory of interaction of irreversible damages. In: Proceedings of V International Symposium on Tribo-Fatigue (ISTF 2005), pp. 10–28. Irkutsk-Bratsk (2005) (in Russian)
5. Kedrov, B.M.: Modern classification of sciences (main trends in its evolution). Dialectics in the science of nature and man. Unity and diversity of the world, differentiation, and integration of scientific knowledge. Moscow (1983) (in Russian)
6. Kiyashchenko, L.P., Moiseev, V.I.: Philosophy of transdisciplinarity. Moscow (2009) (in Russian)
7. Schrődinger, E.: What is Life? From the Point of View of Physics. Atomizdat, Moscow (1972)
8. Sherbakov, S.S., Sosnovskiy, L.A.: Mechanics of Tribo-Fatigue Systems. BSU Press, Minsk (2010). (in Russian)
9. Sherbakov, S.S.: To the general theory of interaction in Tribo-Fatigue systems. Harmonic development of systems is the third way of mankind. In: I International Congress, 8–10 October 2011, pp. 158–163. Proceedings of Institute of Creative Technologies, Odessa (2011)
10. Sosnovskiy, L.A., et al.: Reliability. Risk. Quality. BelSUT Press, Gomel (2012). (in Russian)
11. Sosnovskiy, L.A., et al.: Interaction of sciences. Harmonic development of systems is the third way of humanity: In: Proceedings of 1st International Congress, Odessa, Ukraine, 8–10 October 2011, pp. 115–126, Odessa (2011) (in Russian)
12. Sosnovskiy, L.A., Sherbakov, S.S.: Mechanothermodynamic system and its behavior. Continuum Mech. Thermodyn. **24**(3), 239–256 (2012)
13. Sosnovskiy, L.A., Sherbakov, S.S.: Possibility to construct mechanothermodynamics. Nauka i Innovatsii **60**, 24–29 (2008). (in Russian)
14. Sosnovskiy, L.A., Sherbakov, S.S.: Surprises of Tribo-Fatigue. Magic Book, Minsk (2009)
15. Sosnovskiy, L.A.: Tribo-Fatigue. Wear-Fatigue Damage and Its Prediction (Foundations of Engineering Mechanics). Springer, Berlin (2005)
16. Sosnovskiy, L.A.: Fundamentals of Tribo-Fatigue. BelSUT Press, Gomel (2003). (in Russian)
17. 索斯洛夫斯基著, L.A.: 摩擦疲劳学 磨损 – 疲劳损伤及其预测. 高万振译 – 中国矿业大学出版社 (2013) (in Chinese)
18. Sosnovskiy, L.A.: L-Risk (Mechanothermodynamics of Irreversible Damages). BelSUT Press, Gomel (2004). (in Russian)
19. Sosnovskiy, L.A., Lazarevich, A.A.: Philophosy and Tribo-Fatigue. In: Zhuravkov M.A., et al. (eds.) Proceedings of VI International Symposium on Tribo-Fatigue, Minsk, 25 Oct–1 Nov. 2010, Part 2, pp. 591–620. BSU Press, Minsk (2010)
20. Sosnovskiy, L.A., Makhutov, N.A.: Methodological problems of a comprehensive assessment of damageability and limiting state of active systems. Factory Lab. **5**, 27–40 (1991)

21. Sosnovskiy, L.A., Sherbakov, S.S.: Surpises of Tribo-Fatigue. BelSUT Press, Gomel (2005). (in Russian)
22. Sosnovskiy, L.A., Sherbakov, S.S.: Principles of Mechanothermodynamics. BelSUT Press, Gomel (2013). (in Russian)
23. Sosnovskiy, L.A., Zhuravkov, M.A., Sherbakov, S.S.: Fundamental and Applied Problems of Tribo-Fatigue: A Course of Lectures. BSU Press, Minsk (2010). (in Russian)
24. Sosnovskiy, L.A.: Intelligent dynamic systems: problem and research perspectives. Mechanics 2011. In: Proceedings of V Belarusian Congress on Theoretical and Applied Mechanics, vol. 1, pp. 64–79. Press of the Joint Institute of Mechanical Engineering of NAS of Belarus, Minsk (2011) (in Russian)
25. Sosnovskiy, L.A.: Life field and golden proportions. Nauka i Innovatsii. **79**(9), 26–33 (2009) (in Russian)
26. Sosnovskiy, L.A.: L-risk and golden proportion. Symmetry: art and science. In: Proceedings of International Conference on Harmony in Forms and Processes: Nat. Soc. Sci. Art. **Lvov**, 186–189 (2008)
27. Sosnovskiy, L.A.: Mechanics of Wear-Fatigue Damage. BelSUT Press, Gomel (2007). (in Russian)
28. Sosnovskiy, L.A.: On a possible construction of the general theory of evolution of systems. In: Phylosophy in Belarus and Perspectives of World Intellectual Culture: Proceedings of International Scientific Conference to the 80th Anniversairy of the Institute of Philosophy of NAS Belarus, Minsk, 14–15 April 2011, pp. 152–157 (2011) (in Russian)
29. Sosnovskiy, L.A.: On Dynamic systems with elements of intelligence. In: Zhuravkov M.A., et al. (eds.) Proceedings of VI International Symposium on Tribo-Fatigue, Minsk, 25 Oct–1 Nov 2010, Part 2, pp. 573–582. BSU Press, Minsk (2010) (in Russian)
30. Sosnovskiy, L.A.: On one type of entropy as the absorption measure of energy spent for damage generation in the mechanothermodynamic system. Doklady NAS Belarus **51**(6), 100–104 (2007)
31. Sosnovskiy, L.A.: Statistical Mechanics of Fatigue Damage. Nauka i Tekhnika, Minsk (1987). (in Russian)
32. Sosnovskiy, L.A.: Tribo-Fatigue: About Dialectics of Life. S&P Group TRIBOFATIGUE, Gomel (1999)
33. Sosnovskyi, L.A., et al.: On the interaction of sciences. Harmonic development of systems is the third way of mankind. I International Congress, 8–10 October 2011, pp. 115–126. Proc. Institute of Creative Technologies, Odessa (2011)
34. Zhuravkov, M.A., Sherbakov, S.S.: Interaction in a muticomponent system. In: XI Belarusian Mathematical Conference: Book of Abstracts of the International Scientific Conference, Minsk, BSU, 4–9 Nov 2012, pp. 66–67 (2012) (in Russian)

Chapter 5
Generalized Model of Mechanothermodynamic States of Continuum

Abstract The known models for energy and entropy states of a continuous medium (continuum) are reviewed and analyzed. A generalized model of the mechanothermodynamic state of a medium is developed and analyzed. Its main feature is the rejection of the additivity principle of effective energy (entropy) flows caused by different-nature sources.

5.1 General Notions

In the previous sections different models of energy and entropy states of mechanical, thermodynamic and mechanothermodynamic systems are considered. The construction of the generalized model of a mechanothermodynamic medium will be given below. The analysis and synthesis are based on:

- energy description of continuum (Sect. 5.1);
- entropy description of continuum (Sect. 5.2);
- formulated laws of mechanothermodynamics (Sect. 5.3).

5.2 Energy and Entropy Descriptions of Continuum States

Motion equations for the elementary volume dV are of the form [2, 3]

$$\sigma_{ij,j} + \rho f_i = \rho \dot{v}_i, \quad i = 1, 2, 3, \tag{5.1}$$

where the σ_{ij}'s are the stresses; ρ is the density; the f_i's are volumetric forces; the v_i's are the velocities.

L. Sosnovskiy and S. Sherbakov, *Mechanothermodynamics*,
DOI 10.1007/978-3-319-24981-0_5

The *law of conservation of mechanical energy* for the continuum of the volume V, with regard to the repeated index summation rule is obtained by multiplying scalar Eq. (5.1) by the velocity vector v_i:

$$\int_V v_i \sigma_{ij,j} dV + \int_V \rho v_i f_i dV = \int_V \rho v_i \dot{v}_i dV. \tag{5.2}$$

The right hand-side of Eq. (5.2) is the *change in the kinetic energy* K of the continuum of the volume V:

$$\int_V \rho v_i \dot{v}_i dV = \frac{d}{dt} \int_V \rho \frac{v_i v_i}{2} dV = \frac{d}{dt} \int_V \rho \frac{v^2}{2} dV = \frac{dK}{dt}. \tag{5.3}$$

Based on the known transformations with regard to *Gauss-Ostrogradsky's theorem, the equation for mechnical energy of continuum* [2] is derived

$$\frac{dK}{dt} + \int_V \sigma_{ij} \dot{\varepsilon}_{ij} dV = \int_\Pi \sigma_{ij} l_j d\Pi + \int_V \rho v_i f_i dV, \tag{5.4}$$

or

$$\frac{dK}{dt} + \frac{\delta U}{dt} = \frac{\delta A}{dt},$$

where ε_{ij} is the strain rate; Π is the continuum surface; the l's are the director cones at the continuum surface; $\delta U/dt$ is the power of internal forces; $\delta A/dt$ is the power of internal surfaces and volumetric forces.

The symbol δ in expression (5.4) is used to emphasize that the increment in the general case cannot be an accurate differential.

In *the thermomechanical statement, the rate of change in internal energy* U [2] is usually given by the integral

$$\frac{dU}{dt} = \frac{d}{dt} \int_V \rho u dV = \int_V \rho \dot{u} dV, \tag{5.5}$$

where $u = \lim\limits_{\Delta m \to 0} \frac{u(\Delta m)}{\Delta m}$ is the specific internal energy (internal energy density) of the elementary volume of the mass Δm.

The rate of heat transfer to the continuum is expressed as follows:

$$\frac{\delta Q}{dt} = -\int_\Pi c_i l_i d\Pi + \int_V \rho z dV, \tag{5.6}$$

where c_i is the characteristic of the heat flux per unit area of continuum surface per unit time due to heat conduction; z is the constant of heat radiation per unit mass per unit time.

The *law of change in the energy of thermomechanical continuum* then assumes the form

$$\frac{dK}{dt} + \frac{dU}{dt} = \frac{\delta A}{dt} + \frac{\delta Q}{dt}. \tag{5.7}$$

In (5.7), the transformation of surface integrals into volume integrals allows the *local form of the energy equation* to be obtained

$$\frac{d}{dt}\left(\frac{v^2}{2} + u\right) = \frac{1}{\rho}\left(\sigma_{ij}v_i\right)_{,j} + f_i v_i - \frac{1}{\rho}c_{i,i} + z. \tag{5.8}$$

If the scalar product of Eq. (5.1) and the velocity vector v_i, is subtracted from Eq. (5.8), then the following form of the local energy equation will be obtained:

$$\frac{du}{dt} = \frac{1}{\rho}\sigma_{ij}\dot{\varepsilon}_{ij} - \frac{1}{\rho}c_{i,i} + z = \frac{1}{\rho}\sigma_{ij}\dot{\varepsilon}_{ij} + \frac{dq}{dt}, \tag{5.9}$$

where dq is the heat flux per unit mass.

According to expression (5.9), *the rate of change in the internal energy is equal to the sum of the stress power and the heat flux to the continuum.*

As applied to the *thermodynamic system*, two characteristic functions of its state are defined: absolute temperature T and entropy S that can be interpreted as the characteristic of ordering (or chaotic state) of the thermodynamic system. Usually, it is assumed that the *entropy possesses the property of additivity*, i.e.,

$$S = \sum_i S_i. \tag{5.10}$$

In *continuum mechanics* [2, 3], the *specific entropy per unit mass* is considered

$$S = \int_V \rho s dV. \tag{5.11}$$

The *specific entropy increment* ds can be due to the interaction with the environment (the increment $ds^{(e)}$) or inside the system itself (the increment $ds^{(i)}$) [2, 3]:

$$ds = ds^{(e)} + ds^{(i)}. \tag{5.12}$$

The increment $ds^{(i)}$ is equal to zero in reversible processes and is greater than zero in irreversible processes.

If the heat flux per unit mass is expressed through dq, then for reversible processes the increment will be

$$Tds = dq. \tag{5.13}$$

According to the second law of thermodynamics, the rate of change in the total entropy S of the continuum of the volume V cannot be smaller than the sum of the heat flux through the volume boundary and the entropy produced inside the volume by external sources (*Clasius–Duhem's inequality*) [2, 3]:

$$\frac{d}{dt}\int\limits_V \rho s dV \geq \int\limits_V \rho e dV - \int\limits_\Pi \frac{c_i l_i}{T} d\Pi, \tag{5.14}$$

where e is the power of local external entropy sources per unit mass. The equality in formula (5.14) is valid for reversible processes and the inequality—irreversible processes.

Transforming the surface integral into the volume integral in expression (5.14) can yield the relation for the *rate of the internal entropy production per unit mass*:

$$\gamma \equiv \frac{ds}{dt} - e - \frac{1}{\rho}\left(\frac{c_i}{T}\right)_{,j} \geq 0. \tag{5.15}$$

In continuum mechanics, it is assumed that the stress tensor can be decomposed into two parts: the *conservative part* $\sigma_{ij}^{(C)}$ for reversible processes (elastic deformation, liquid pressure) and the *dissipative part* $\sigma_{ij}^{(D)}$ for irreversible processes (plastic deformation, liquid viscous stresses):

$$\sigma_{ij} = \sigma_{ij}^{(C)} + \sigma_{ij}^{(D)}. \tag{5.16}$$

The expression for the *rate of change in energy* (5.9) can then be presented in the following form:

$$\frac{du}{dt} = \frac{1}{\rho}\sigma_{ij}\dot{\varepsilon}_{ij} + \frac{dq}{dt} = \frac{1}{\rho}\sigma_{ij}^{(C)}\dot{\varepsilon}_{ij} + \frac{1}{\rho}\sigma_{ij}^{(D)}\dot{\varepsilon}_{ij} + \frac{dq}{dt}. \tag{5.17}$$

If it is assumed that relation (5.13) is valid for irreversible processes, then the *rate of the total entropy production* is

$$\frac{ds}{dt} = \frac{1}{\rho T}\sigma_{ij}^{(C)}\dot{\varepsilon}_{ij} + \frac{1}{\rho T}\sigma_{ij}^{(D)}\dot{\varepsilon}_{ij} + \frac{1}{T}\frac{dq}{dt}, \tag{5.18}$$

or

$$\frac{ds}{dt} = \frac{1}{\rho T}\left(\frac{du_M}{dt} + \frac{du_T}{dt}\right) = \frac{1}{\rho T}\left(\frac{du_M^{(C)}}{dt} + \frac{du_M^{(D)}}{dt} + \frac{du_T}{dt}\right).$$

Expression (5.18) for the rate of a total change of local entropy in the elementary volume of the continuum can be very convenient in practice.

In view of assumption (5.10) on the entropy additivity, the sum in (5.18) can be supplemented with other terms considering the internal entropy production in the liquid (gas) volume due to different mechanisms. Similarly, for the continuum volume dV, for example, internal chemical processes can be allowed for in formulas (2.2), (2.3) [1].

If dV is considered not as a finite volume but as an *elementary volume of continuum*, then changes in its specific energy and entropy based on (5.17), (2.2) and (2.3) can be written in the following differential form:

$$du = \frac{1}{\rho}\sigma_{ij} d\varepsilon_{ij} + dq + \sum_k \mu_k dn_k; \tag{5.19}$$

$$ds = \frac{1}{\rho T}\sigma_{ij} d\varepsilon_{ij} + \frac{1}{T} dq + \frac{1}{T}\sum_k \mu_k dn_k, \tag{5.20}$$

where n_k is the number of mols per unit mass.

For the *continuum of the volume V*, expressions (5.19) and (5.20) on the basis of relations (5.5) and (5.11) will assume the form

$$dU = \int_V \rho du dV = \int_V \sigma_{ij} d\varepsilon_{ij} dV + \int_V \rho dq dV + \int_V \rho \sum_k \mu_k dn_k dV; \tag{5.21}$$

$$dS = \int_V \rho ds dV = \frac{1}{T}\int_V \sigma_{ij} d\varepsilon_{ij} dV + \frac{1}{T}\int_V \rho dq dV + \frac{1}{T}\int_V \rho \sum_k \mu_k dn_k dV. \tag{5.22}$$

The introduction of the *chemical entropy component* [the last terms in (5.19–5.21)] allowed one not only to obtain a more full picture of the continuum state but also to describe self-organization processes that result in initiating stable structures when the heat flux to the continuum is increased.

The above-presented known models of energy and entropy states of continuum (5.17–5.22), being rather general, do not nevertheless permit one to satisfactorily describe some processes occurring in such a continuum as a deformable solid. However a convenient representation of the additivity of energy and entropy components (5.11) used, for example, for modeling elastic deformation is not suitable for the description of non-linear processes. The available models do not also allow for the entropy growth due to damageability of solids as a specific characteristic of changes in the structural organization. According to the Tribo-Fatigue concepts [4, 9], the damageability is interpreted as any irreversible change in structure, continuum, shape, etc. of a deformable solid that leads to its limiting state. Although, for example, the elasticity limit is not taken into account implicitly during plasticity modeling, the damageability, for example, during mechanical or contact fatigue occurs in the conditions of linear elastic deformation and requires a particular approach for its description with regard to limiting fatigue characteristics of material. The above drawbacks are overcome in the below approach.

5.3 Mechanothermodynamic States of Continuum

Within the framework of Mechanothermodynamics, a special approach is being developed to assess the entropy in terms of a generalized energy state. Following this approach and formula (2.52), out of the *total energy* (specific) due to some particular loads (force, temperature, etc.), *its effective part directly spent for the damage production is defined* by the experimentally found coefficients A_l [4, 9]

$$u_l^{eff} = A_l u_l, \tag{5.23}$$

where the u_l 's are the specific internal energies at tear (u_n), shear (u_T), thermal action (u_T).

Total specific energy of an elementary volume and a *rate of its change* are then given as

$$u = \sum_l \left[(1 - A_l) u_l + u_l^{eff} \right]; \tag{5.24}$$

$$\frac{du}{dt} = \sum_l \left[(1 - A_l) \frac{du_l}{dt} + \frac{du_l^{eff}}{dt} \right]. \tag{5.25}$$

Moreover, the Λ-functions are used to take into consideration a complex (nonadditive) character of interactions between effective energies of different nature expressed by formula (2.18). This allows the *total effective energy of the system* to be assessed:

$$u_\Sigma^{eff} = \Lambda_\alpha \left(u_l^{eff} \right) = \Lambda_{M \backslash T} \left(\Lambda_{\tau \backslash n}, A_l u_l \right) = \Lambda_{M \backslash T} \left\{ \Lambda_{\tau \backslash n} [A_n u_n + A_\tau u_\tau] + A_T u_T \right\}, \tag{5.26}$$

where the Λ_α's are the possible combinations of interaction of effective energies (irreversible damages).

The *specific feature of* Λ-functions is such that

$$u_{\Sigma}^{eff} \gtreqless u_l^{eff}, \tag{5.27}$$

and, hence,

$$u_{\Sigma}^{eff} \gtreqless \sum u. \tag{5.28}$$

Thus, using coefficients A_l and Λ-functions, it is possible to assess *energy interaction* due to different-nature loads. Such interaction can cause both a sharp growth and a substantial decrease of effective energy, resulting in damages and limiting states, as compared to the one calculated by the ordinary additivity model of type (5.17):

$$u_{\Sigma} = \sum A_l u_l. \tag{5.29}$$

The *total effective energy* of the volume V and its accumulation in time with regard to formula (5.26) are of the form

$$U_{\Sigma}^{eff} = \int_V \rho u_{\Sigma}^{eff}(V)dV \tag{5.30}$$

and

$$U_{\Sigma}^{eff}(t) = \int_t \int_V \rho u_{\Sigma}^{eff}(V,t)dVdt. \tag{5.31}$$

The principal moment of the mechanothermodynamic model is the *account of the limiting state* (limits of plasticity, strength, fatigue, etc.) according to Item XIII (page 12)

$$u_{\Sigma}^{eff} = u_0, \tag{5.32}$$

where u_0 is the limiting density of the internal energy interpreted as the initial activation energy of the disintegration process.

A relationship between the current state (mechanical, thermomechanical, energy) of an elementary volume of a solid (medium) and its limiting state enables one to construct the *parameter of local energy damageability*: dimensionless

$$\psi_u^{eff} = \frac{u_{\Sigma}^{eff}}{u_0} \tag{5.33}$$

or dimensional

$$\psi_{u^*}^{eff} = u_{\Sigma}^{eff} - u_0. \tag{5.34}$$

Local energy damageability (5.33) or (5.34) is most general out of the damageability parameters constructed in terms of different mechanical (thermomechanical) states φ [4, 9]:

$$\psi_q = \varphi_q / \varphi_q^{(*\lim)}, \tag{5.35}$$

where $\varphi = \sigma, \varepsilon, u$; the σ's are the stresses; the ε's are the strains; u is the density of internal energy; the $\varphi_q^{(*\lim)}$'s are the limiting values of the state $\varphi \in \left\{\text{eqv}, ij, i, S, {}^{D}_{ij}, n, \tau, int, u, {}^{n}_{u}, {}^{\tau}_{u}, {}^{eff}_{u}\right\}$; eqv is the equivalent mechanical state; the ij 's are the components of the tensor φ; the i 's are the main components of the tensor φ; S and ${}^{D}_{ij}$ and the sphere and deviator parts of the tensor φ; n and τ are the normal and tangential components of the tensor φ; int is the intensity φ; u is the specific potential strain energy (internal energy density); the indices at u mean: ${}^{n}_{u}$ and ${}^{\tau}_{u}$ are the specific potential strain energy at tension–compression and shear; ${}^{eff}_{u}$ is the effective specific potential strain energy.

Integral damageability measures can be built on the basis of local measures (5.35) *with the use of the model of a deformable solid with a dangerous volume* [10].

By the *dangerous volume* is understood the spatial region of a loaded solid, at each point of which the value of local damageability is smaller than the limiting one [4, 9]:

$$V_q = \left\{ dV / \varphi_q \geq \varphi_q^{(*\lim)}, \quad dV \subset V_k \right\}, \tag{5.36}$$

or

$$V_q = \left\{ dV / \psi_q \geq 1, \quad dV \subset V_k \right\}.$$

Dangeous volumes are calculated by the following general formula:

$$V_q = \iiint\limits_{\psi_q(V) \geq 1} dV. \tag{5.37}$$

The *integral condition of damageability* of a solid or a system can be written in the form

$$0 < \omega_q = \frac{V_q}{V_0} < 1, \tag{5.38}$$

where V_0 is the working volume of the solid.

To analyze at a time dangerous volumes local damageability distributed within them, the *function of damageability of unit volume* is introduced

$$d\Psi_q = \psi_q(V)dV. \tag{5.39}$$

The *function of damageability of the entire volume V* will then be as

$$\Psi_q = \int\limits_{\psi_q \geq 1} \psi_q(V)dV. \tag{5.40}$$

The simplest functions of *damageability accumulation in time* for the unit volume and the entire volume will be of the following form, respectively

$$d\Psi_q^{(t)} = \int\limits_t \psi_q(t)dt; \tag{5.41}$$

$$\Psi_q^{(t)} = \int\limits_{\psi_q \geq 1} \int\limits_t \psi_q(V,t)dt\,dV. \tag{5.42}$$

The *indices of volume-mean damageability*

$$\bar{\Psi}_q^{(V)} = \frac{1}{V_q} \int\limits_{\psi_q \geq 1} \psi_q(V)dV \tag{5.43}$$

and *its accumulation in time* can be used

$$\bar{\Psi}_q^{(V,t)} = \frac{1}{V_q} \int\limits_t \int\limits_{\psi_q \geq 1} \psi_q(V,t)dV\,dt \tag{5.44}$$

The analysis of formulas (5.33), (5.39) and (5.41) comes to the conclusion that conceptually, they are related to the concept of entropy as a difference (or relations) between two states (configurations) of a system, the degree of its organization (chaotic state). As applied to damageability, such states are current and limiting.

Now using local energy damageability (5.33), construct *specific* (per unit mass) *Tribo-Fatigue entropy* (up to a constant):

$$s_{TF} = \psi_u^{eff}\left(\Lambda_\alpha, A_l, \sigma_{ij}, T\right) = \lim_{\Delta m \to 0} A_\psi \frac{u_\Sigma^{eff}(\Delta m)}{u_0 \Delta m}, \tag{5.45}$$

or

$$s_{TF} = s_{TF^*} = \frac{\psi_{u^*}^{eff}(\Lambda_\alpha, A_l, \sigma_{ij}, T)}{T} = \frac{u_\Sigma^{eff} - u_0}{T}. \tag{5.46}$$

where A_ψ is the dimensional parameter (J mol^{-1} K^{-1}).

On the basis of the expressions for entropy (5.18), as well as of formulas (5.24), (5.25) the *local entropy* and the *rate of its change within an elementary volume* will be

$$s = \frac{1}{T}\sum_l [(1 - A_l)u_l] + s_{TF} \tag{5.47}$$

and

$$\frac{ds}{dt} = \frac{1}{T}\sum_l \left[(1 - A_l)\frac{du_l}{dt}\right] + \frac{ds_{TF}}{dt}. \tag{5.48}$$

From formulas (5.47) and (5.48) it is seen that unlike the thermomechanical model, the *state indices of the mechanothermodynamic system u* and *s are not equivalent*. This is due to the fact that the calculation of the Tribo-Fatigue entropy s_{TF} by formula (5.45) is supplemented with the limiting state in the form of the limiting density of the internal energy u_0.

The Tribo-Fatigue entropy S_{TF} is calculated not within the entire volume *V*, but only within its damageable part, i.e., within the energy effective dangerous volume V_u^{eff}:

$$V_u^{eff} = \left\{dV/u_\Sigma^{eff} \geq u_0, \quad dV \subset V_k\right\}. \tag{5.49}$$

On the basis of formulas (5.11), (5.45) and (5.49), the *Tribo-Fatigue entropy of volume V* will be

$$S_{TF} = \int\limits_{u_\Sigma^{eff}(V) \geq u_0} \rho s_{TF}(V) dV = \int\limits_{u_\Sigma^{eff}(V) \geq u_0} \rho \psi_u^{eff}(V) dV, \tag{5.50}$$

where

$$\psi_u^{eff}(V) = \frac{u_\Sigma^{eff}(V)}{u_0} \text{ or } \psi_u^{eff}(V) = \frac{\psi_{u^*}^{eff}(V)}{T} = \frac{u_\Sigma^{eff}(V) - u_0}{T(V)}, \tag{5.51}$$

and *its accumulation* will be

$$S_{TF}(t) = \int_t \int_{u_\Sigma^{eff}(V,t) \geq u_0} \rho s_{TF}(V,t) dV\, dt = \int_t \int_{u_\Sigma^{eff}(V,t) \geq u_0} \rho \psi_u^{eff}(V,t) dV dt, \quad (5.52)$$

where

$$\psi_u^{eff}(V,t) = \frac{u_\Sigma^{eff}(V,t)}{u_0} \text{ or } \psi_u^{eff}(V,t) = \frac{\psi_{u^*}^{eff}(V,t)}{T(V,t)} = \frac{u_\Sigma^{eff}(V,t) - u_0}{T(V,t)}. \quad (5.53)$$

The *principal feaure of Tribo-Fatigue total* S_{TF} and specific s_{TF} entropies should be emphasized. They allow the difference between two states to be assessed not only quantitatively (as thermomechanical entropy), but also qualitatively, as the value of the limiting density of the internal energy u_0 is explicitly introduced into the calculation of the specific entropy s_{TF}. Thus, s_{TF} and S_{TF} allow one to answer the question how much the current state of a solid or a system is dangerous in comparison with limiting states.

The *total entropy* and the *rate of its change for a solid of a system* with regard to (5.50) and (5.52) assume the form

$$S = \int_V \frac{1}{T(V)} \sum_l \rho[(1 - A_l(V))u_l(V)] dV + S_{TF} \quad (5.54)$$

and

$$\frac{dS}{dt} = \int_V \frac{1}{T(V)} \sum_l \rho\left[(1 - A_l(V))\frac{du_l(V)}{dt}\right] dV + \frac{dS_{TF}}{dt}. \quad (5.55)$$

Based on formulas (5.45)–(5.55), the *function of accumulation of total entropy in time* can be built:

$$\begin{aligned} S(t) &= \int_t \int_V \sum_l \rho s_l(V,t) dV dt + \int_t \int_{u_\Sigma^{eff}(V,t) \geq u_0} \rho s_{TF}(V,t) dV dt \\ &= \int_t \int_V \frac{1}{T(V,t)} \sum_l \rho\left[(1 - A_l(V,t))\frac{du_l(V,t)}{dt}\right] dV\, dt + \int_t \int_{u_\Sigma^{eff}(V,t) \geq u_0} \rho \psi_u^{eff}(V,t) dV dt. \end{aligned}$$

(5.56)

In this respect, bearing in mind the limiting states of a solid or a system, models (5.54)–(5.56) permit one to answer the question whether the current state is the point of a qualitative jump in the system, i.e., whether the current state is close to the limiting one. A similar (*dialectic as a matter of fact*) *qualitative transition* differs, for example, from the bifurcation point having the uncertainty in a further development of events and the possibility to predict the system behavior after a transition on the basis of the analysis of s_{TF} and S_{TF}. Particular limiting states (limit of strength, mechanical or contact fatigue, etc.) enable one to predict the situation after transiting the given point: principal changes in the system properties and behavior or the formation of a new system based on the previous one.

As an example, there can be non-linear deformation or generation of microcracks in the solid (or the system) that cause the changes in its strength and fatigue properties, and, hence, to its response to loads. In turn, formed macrocracks lead to local continuum violation—formation of new free surfaces (possibly, of new solids—destruction products), i.e. a new system.

It should be noted that models (5.54)–(5.56) were built using a traditional concept of entropy additivity (5.10) although with regard to substational improvements. These models also contain reversible processes described by the entropy components s_l not yielding primary damages and, hence, the limiting states—the points of a qualitative change of the system.

It is therefore more advisable for a qualitative and quantitative analysis of evolution of systems (whose states are traditionally defined as bifurcation branches) that the entropy state should be determined by means of the mechanothermodynamic model of the solid using only Tribo-Fatigue entropy. In this case, formulas (5.50)–(5.52) for entropy and its accumulation will be of the form

$$S = S_{TF} = \int\limits_{u_\Sigma^{eff}(V,t) \geq u_0} \rho s_{TF}(V)dV = \int\limits_{u_\Sigma^{eff}(V,t) \geq u_0} \rho \psi_u^{eff}(V)dV, \tag{5.57}$$

and

$$S(t) = S_{TF}(t) = \int\limits_t \int\limits_{u_\Sigma^{eff}(V,t) \geq u_0} \rho s_{TF}(V)dVdt = \int\limits_t \int\limits_{u_\Sigma^{eff}(V,t) \geq u_0} \rho \psi_u^{eff}(V,t)dVdt. \tag{5.58}$$

To identify the points of qualitative change in the limiting states of solids (systems), the indices of relative integral entropy and its accumulation built on the basis of the concept of integral conditioned of solid damage (2.19) can be used (5.38):

$$\omega_S = \frac{S_{TF}}{V_0} = \frac{1}{V_0} \int\limits_{u_\Sigma^{eff}(V,t) \geq u_0} \rho s_{TF}(V)dV; \tag{5.59}$$

$$\omega_S(t) = \frac{S_{TF}(t)}{V_0} = \frac{1}{V_0} \int\limits_t \int\limits_{u_\Sigma^{eff}(V,t) \geq u_0} \rho s_{TF}(V)dVdt. \tag{5.60}$$

The indices S_{TF}, $S_{TF}(t)$, ω_S, $\omega_S(t)$ *can grow infinitely, allowing not only the limiting states of type* (5.32), *but also different transmitting states to be described;* in essence, they "provide" a quantitative description of the *law of increase of entropy.*

Now, based on formulas (5.22), (5.54), (5.56) and (5.58), let us construct *generalized expresssions for entropy, a rate of its change,* as well as *its accumulation in the mechanothermodynamic system consisting of a liquid (gas) medium of the volume V* and *a solid of the volume V_ψ*:

$$\begin{aligned} S &= \int\limits_V \rho s_T dV + \int\limits_{V_\psi} \sum_l \rho s_l dV_\psi + \int\limits_{u_\Sigma^{eff} \geq u_0} \rho s_{TF} dV_\psi \\ &= \int\limits_V \frac{1}{T}\sigma_{ij}\varepsilon_{ij} dV + \int\limits_V \frac{1}{T}\rho q dV + \int\limits_V \frac{1}{T}\rho \sum_k \mu_k n_k \, dV \\ &+ \int\limits_{V_\psi} \frac{1}{T}\sum_k \rho[(1-a_k)u_k] dV_\psi + \int\limits_{u_\Sigma^{eff} \geq u_0} \rho\psi_u^{eff} dV_\psi; \end{aligned} \tag{5.61}$$

$$\begin{aligned} \frac{dS}{dt} &= \int\limits_V \rho\frac{ds_T}{dt} dV + \int\limits_{V_\psi} \sum_l \rho \frac{ds_l}{dt} dV_\psi + \int\limits_{V_\psi} \rho\frac{ds_{TF}}{dt} dV_\psi \\ &= \int\limits_V \frac{1}{T}\sigma_{ij}\frac{d\varepsilon_{ij}}{dt} dV + \int\limits_V \frac{1}{T}\rho\frac{dq}{dt} dV + \int\limits_V \frac{1}{T}\rho\sum_k \mu_k \frac{dn_k}{dt} dV \\ &+ \int\limits_{V_\psi} \frac{1}{T}\sum_k \rho\left[(1-a_k)\frac{du_k}{dt}\right] dV_\psi + \int\limits_{u_\Sigma^{eff} \geq u_0} \rho\frac{d\psi_u^{eff}}{dt} dV_\psi; \end{aligned} \tag{5.62}$$

$$\begin{aligned} S(t) &= \int\limits_t \int\limits_V \rho s_T dV\, dt + \int\limits_t \int\limits_{V_\psi} \sum_l \rho s_l dV_\psi \, dt + \int\limits_t \int\limits_{u_\Sigma^{eff} \geq u_0} \rho s_{TF} dV_\psi dt \\ &= \int\limits_t \int\limits_V \frac{1}{T}\sigma_{ij}\varepsilon_{ij} dV dt + \int\limits_t \int\limits_V \frac{1}{T}\rho q dV\, dt + \int\limits_t \int\limits_V \frac{1}{T}\rho\sum_k \mu_k n_k dV\, dt \\ &+ \int\limits_t \int\limits_{V_\psi} \frac{1}{T}\sum_l \rho[(1-a_l)u_l] dV_\psi dt + \int\limits_t \int\limits_{u_\Sigma^{eff} \geq u_0} \rho\psi_u^{eff} dV_\psi dt. \end{aligned} \tag{5.63}$$

Similarly, entropy state indices can be built for a system composed of many media.

It should be noted that the interaction (contact) of two media in formulas (5.61)–(5.64), which can be complex in nature, is taken into account only implicitly in terms of medium state parameters (stresses, strains, temperature). It is obvious that this is only the first step to a comprehensive (generalized) solution of the problem stated.

The simplified writing of expression (5.62) for the entropy increment of the mechanothermodynamic system composed of finite volumes dV and dV_ψ was given in [6, 8] in the form of formula (3.11) that can be re-written in the following form:

$$dS = (dS)_T + (d_i S)_{TF} = \frac{dU + \Delta p dV}{T} - \frac{1}{T}\sum_k \mu_k dN_k + \Psi_u^{eff} dV_\psi. \tag{5.64}$$

Expression (5.64) can also be represented in terms of specific quantities as:

$$dS = \int_V \frac{\rho du + \rho dp}{T} dV - \int_V \frac{1}{T}\rho \sum_k \mu_k dn_k dV + \int_{u_\Sigma^{eff} \geq u_0} \rho d\psi_u^{eff} dV_\psi \tag{5.65}$$

or on the basis of (5.62) as:

$$\frac{dS}{dt} = \int_V \frac{\sigma_{ij} d\varepsilon_{ij} + \rho dq}{T dt} dV - \int_V \frac{1}{T}\rho \sum_k \mu_k \frac{dn_k}{dt} dV + \int_{u_\Sigma^{eff} \geq u_0} \rho \frac{d\psi_u^{eff}}{dt} dV_\psi. \tag{5.66}$$

In formulas (5.50)–(5.52) for calculation of the Tribo-Fatigue entropt S_{TF} and its accumulation $S_{TF}(t)$, the specific entropy s_{TF} is assumed to be integrated in terms only of the damageable part of the solid—the dangerous volume. However the influence of undamagable regions can also be allowed for by integrating S_{TF} over the entire volume:

$$S_{TF} = \int_V \rho s_{TF}(V) dV = \int_V \rho \psi_u^{eff}(V) dV; \tag{5.67}$$

$$S_{TF}(t) = \int_t \int_V \rho s_{TF}(V,t) dV dt = \int_t \int_V \rho \psi_u^{eff}(V,t) dV dt, \tag{5.68}$$

where

$$\psi_u^{eff} = \begin{cases} A_\psi \frac{u_\Sigma^{eff}(V,t)}{u_0} \geq 1, & \text{if } u_\Sigma^{eff} \geq u_0; \\ A_\psi \frac{u_\Sigma^{eff}(V,t)}{u_0} < 1, & \text{if } u_\Sigma^{eff} < u_0, \end{cases} \tag{5.69}$$

or

$$\psi_u^{eff} = \frac{\psi_{u^*}^{eff}(V,t)}{T(V,t)} = \begin{cases} \frac{u_\Sigma^{eff}(V,t) - u_0}{T(V,t)} \geq 0, & \text{if } u_\Sigma^{eff} \geq u_0; \\ \frac{u_\Sigma^{eff}(V,t) - u_0}{T(V,t)} < 0, & \text{if } u_\Sigma^{eff} < u_0. \end{cases} \tag{5.70}$$

From expression (5.70) it is seen that $\psi_u^{eff} < 0$ is observed outside the dangerous volume (at $u_\Sigma^{eff} < u_0$). This means that the *specific Tribo-Fatigue entropy s_{TF} also appears to be negative* (or less than unity for its alternative definition) *outside the* *dangerous volume* where the limiting state is not reached. Negative values of ψ_u^{eff} and s_{TF} can then be interpreted as the absence of damageability or, in other words, as the retention of structure and/or properties of the solid.

As follows from the above-stated, the assumption on the entropy additivity is wrong in the general case for a system composed of both a solid and a liquid (gas) where chemical reactions can occur. By analogy with Λ-functions, interaction functions of different-nature energy (5.26), it is necessary to introduce *interaction functions of different-nature entropy* by adding them to expression (5.64) in effort to determine total effective entropy:

$$\begin{aligned} dS_{total}^{eff} &= \Lambda_{T\backslash TF}^{(S)}(dS_T + d_i S_{TF}) = \Lambda_{T\backslash TF}^{(S)}\left[\Lambda_{Q\backslash Ch}^{(S)}(dS_T^Q + dS_{Ch}^Q) + d_i S_{TF}\right] \\ &= \Lambda_{T\backslash TF}^{(S)}\left[\Lambda_{Q\backslash Ch}^{(S)}\left(\frac{dU + \Delta p dV}{T} - \frac{1}{T}\sum_k \mu_k dN_k\right) + \Psi_u^{eff} dV_\psi\right], \end{aligned} \tag{5.71}$$

or

$$\begin{aligned} dS_{total}^{eff} &= \Lambda_{T\backslash TF\backslash Ch}^{(S)}(dS_T + d_i S_{TF}) \\ &= \Lambda_{T\backslash TF\backslash Ch}^{(S)}\left[\frac{dU + \Delta p dV}{T} - \frac{1}{T}\sum_k \mu_k dN_k + \Psi_u^{eff} dV_\psi\right], \end{aligned} \tag{5.72}$$

where the subscripts Q and Ch denote the thermodynamic and chemical entropy components.

Formulas (5.71)–(5.72) are supplemented with the *generalized interaction functions* $\Lambda_{T\backslash TF}^{(S)}$, $\Lambda_{Q\backslash Ch}^{(S)}$, $\Lambda_{T\backslash TF\backslash Ch}^{(S)}$ *in mechanothermodynamic systems*. This means that the *hypothesis on the additivity of thermodynamic and Tribo-Fatigue entropy is not accepted*. The appropriate Λ-functions, interaction functions should then be specified and introduced into Eqs. (5.71) and (5.72).

At last, the following interrelations of damages (5.33), entropy (5.45) and information (with regard to Fig. 2.24) can be obtained:

$$0 \le \psi_u^{eff} = u_\Sigma^{eff} / u_0, \quad s_{TF} = a_s u_\Sigma^{eff} / u_0 \le \text{\textcrh}, \tag{5.73}$$

$$\left.\begin{aligned} 0 \le I = a_I u_\Sigma^{eff} \psi_u^{eff} / u_0 \le 1, \quad &\text{если} \quad u_\Sigma^{eff} \le u_0, \\ -1 \le I^* = -a_I^* u_\Sigma^{eff} \psi_u^{eff} / u_0 \le -\text{\textcrh}, \quad &\text{если} \quad u_\Sigma^{eff} > u_0. \end{aligned}\right\} \tag{5.74}$$

Here, it is assumed that the source (reason) of changes in entropy and information in the system is not associated with energy but with physical damages due to it

$$u_\Sigma^{eff} \to \psi_u^{eff} \to S_{TF} < \begin{matrix} (+I) \\ (-I^*) \end{matrix} \tag{5.75}$$

Thus, the models of energy, entropy, and information states of mechanothermodynamic systems as well as their evolution systems have been described above.

The *central concept for all these models* is *damage*—any irreversible change in sizes, properties and states of the solid or medium. *The distinctive feature of the model as presented here is that the absorbed part of energy spent for damage production is released out of the total energy supplied to the system*. Moreover, it is taken into account that different-nature energies (force, temperature, etc. are not summed up and interact in an intricate manner. Generation and motion of damages are peculiar for any type of loading; their non-linear interaction and accumulation results in the fact that effective energy achieves its appropriate corresponding limiting value.

The model of Tribo-Fatigue entropy as a function of damageability of the solid or the system is very beneficial in effort to describe the points of qualitative dialectic transformations of the system, i.e., *its evolution*. *The value and character of damageability enable predicting not only the moment of arising these transformations but also their direction*. This allows an uncertainty in a further fate of the system to be removed when these points are interpreted as bifurcation ones. Of importance is that *the information associated with damageability is usually interpreted as qualitative*: *positive information is caused by the process of development, whereas negative information—by the process of degradation*.

Such interrelation of damages (entropy) and information makes it possible take a new view of the evolution of living and non-living systems and the life as a whole as special means of accumulation of irreversible damages [7].

References

1. Kondepudi, D., Prigogine, I.: Modern Thermodynamics. From Heat Engines to Dissipative Structures. John Wiley & Sons, Chichester (1998)
2. Mase, G.: Theory and Problems of Continuum Mechanics. McGraw-Hill, New York (1970)
3. Sedov, L.I.: Mechanics of a Continuous Medium. Nauka, Moscow (1973). (in Russian)
4. Sherbakov, S.S., Sosnovskiy, L.A.: Mechanics of Tribo-Fatigue Systems. BSU Press, Minsk (2010). (in Russian)
5. Sosnovskiy, L.A., et al.: Reliability. Risk. Quality. BelSUT Press, Gomel (2012). (in Russian)
6. Sosnovskiy, L.A., Sherbakov, S.S.: Mechanothermodynamic system and its behavior. Continuum Mech. Thermodyn. **24**(3), 239–256 (2012)
7. Sosnovskiy, L.A., Lazarevich, A.A.: Philophosy and Tribo-Fatigue. In: Zhuravkov M.A., et al. (eds.) Proceedings VI International Symposium on Tribo-Fatigue, 25 Oct–1 Nov 2010. Part 2, pp. 591–620. BSU Press, Minsk (2010)
8. Sosnovskiy, L.A., Sherbakov, S.S.: Surpises of Tribo-Fatigue. BelSUT Press, Gomel (2005). (in Russian)
9. Sosnovskiy, L.A.: Mechanics of Wear-Fatigue Damage. BelSUT Press, Gomel (2007). (in Russian)
10. Sosnovskiy, L.A.: Statistical Mechanics of Fatigue Damage. Nauka i Tekhnika, Minsk (1987). (in Russian)

Chapter 6
Some Areas for Further Research

States of MTD system. *Possible physical states of an MTD system are described with the use of appropriate parameters*. Figure 4.17 demonstrates the information (to a first approximation) of different states of the MTD system and its governing parameters. Although these states seem to be very different, it was shown that all they are in principle determined by the energy U, its effective component $U_{\Sigma}^{eff} < U$, the limiting value $U_{\lim}$.

It is not difficult to see that finally, any of nine states shown in Fig. 4.17 are in general described in terms of *damageability* (Fig. 4.17). Such a representation is, hence, *fundamental* since as it has been mentioned in the monograph, damages are observed and measured physically. Characteristic stages of evolution of the MTD system can be established and described in terms of the damageability character. That is why, the *development of the generalized theory of A-evolution* (Fig. 4.17) in its all diverse aspects should be considered an urgent and useful task. Presented here are the basic references that contain the initial analysis of *states* ①… ⑩ according to Fig. 4.17 as applied to Tribo-Fatigue and mechanothermodynamic systems: stress-strain state [6]; energy state [20]; thermodynamic state [10, 11, 20]; limiting state [6, 20]; damageability state [19, 20, 25]; reliability state [8, 20]; entropy state [10, 11, 22]; information state [17, 20, 21]; risk/safety state [8, 13, 15]; transmitting state [16, 20].

Dynamic system with some elements of the intelligence. Now, returning to Fig. 1.1, it is seen that the figure is ended with the arrow asking: *what object will be there, behind the mechanothermodynamic system*? The answer is obvious and general: *the real world* around us. Nowadays it is being intensively studied by many and diverse sciences—from chemistry to biology … through mechanics and thermodynamics … and up to philosophy; from all points of view. From Fig. 1.1 it is also clear that the mechanothermodynamic system must be followed by a little more complex object; however, the object must be much simpler than the real system, for example, *the mechanothermodynamic system with some "elements of" intelligence*.

The first works in this field of research are known [13, 18, etc.].

L. Sosnovskiy and S. Sherbakov, *Mechanothermodynamics*,
DOI 10.1007/978-3-319-24981-0_6

Figure 6.1 is a possible addition to Fig. 1.1—*biomechanics* [1, 26] and *homomechanics* [2, 12, 15, 16, 19, 23]. Of course, it is clear that the *study of such systems* is an actual and perspective *task*.

Information and damageability. But… *where from are the elements of intelligence? This question is also addressed in the book.* Approaches to answering the question lie in the research of *the* Tribo-*Fatigue triada*: *dynamics* → *damage* → *information*. Thus, studies of the processes of *information accumulation and transformation in the moving and damageable system are a task* that is urgent and interesting. The first proposals to solve it can be found in [20, 21]. But it is necessary to find a fundamental solution for the assessment of the *information quality* and then the *generalized quality* of systems [8, 20], including the one with the "elements of intelligence."

Risk and safety. The modern society is seriously concerned with the problems of *risk* and *safety* of existence of the human civilization itself due to an increased growth of man-made risks in combination with natural cataclysms and catastrophes. It is therefore difficult to overestimate the importance of the *task of assessment and control* of the risk of the MTD system; the objective is to provide *its safety* [3, 4, 15].

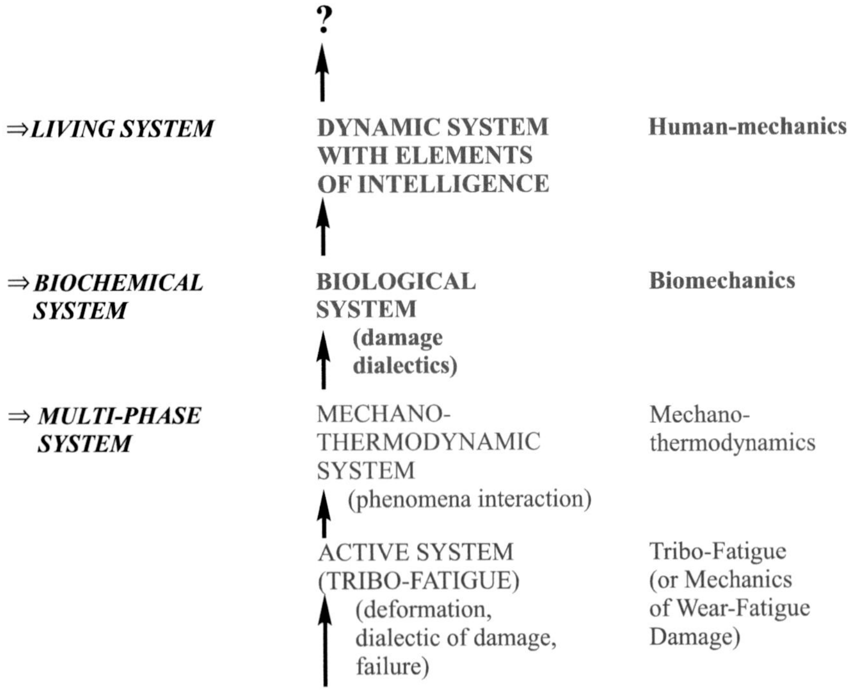

Fig. 6.1 Main directions for further research

Problem of interaction. Diverse real interactions of particles, objects, systems are being studied for a long time and successfully, *especially in physics and philosophy. Fundamental laws* of their interaction and, hence, mutual influences are established. It should be noted: usually, we are talking about *external interactions*, i.e., about the behavior of particles, systems under the action of external (relative to them) forces, energies, fields, etc.

Let us remind that the traditional studies of interactions of physical damages moving in the solid are also being made for a long time and successfully. But as a rule, in the known works the hypothesis on damage accumulation at $\Lambda = 1$ is assumed; but there according to our terminology, the *dialectics of damages*, i.e., the situations when $\Lambda \gtreqless 1$ is not considered.

We have introduced the concept of *internal—dialectic interactions in a moving and deformable solid*. These are caused by the so-called Λ-*interaction of absorbed parts of energy and, hence, effective damages due to many different-nature sources* (mechanical loads, thermodynamic fields, etc.). This gave an impetus to the development of the methods of *phenomenon analysis in Tribo-Fatigue* [20] *and further in mechanothermodynamics*—in addition to the traditional *factor analysis in mechanics*. The development of procedures of phenomenon analysis is the urgent task.

The development of *general theory of* Λ-*interactions* is of great practical importance, since just Λ-interactions define the *fate of the system* (its *evolution*). In [10, 11, 20], the *problem of interaction of particular entropy components*—thermodynamic, Tribo-Fatigue, and chemical (the parameter $\Lambda^{(S)}_{T/TF/Ch}$) was stated for the first time. In particular, it was aimed at revealing and describing quantitatively entropy-critical (or limiting) states of systems. Its development is waiting for its researchers.

Nonclassical tasks. As follows from the above-said, mechanothermodynamics initiates the development of new models and tasks: Tribo-Fatigue entropy [20, 22]; nonclassical model of a deformable solid [9]; nonclassical statement of the dynamic task [17]; phenomenon analysis [20]; dialectic interactions of damages, effective energies, dangerous volumes [7, 20]; nonclassical statement of reliability tasks [14]; nontraditional concept of risk (safety) [13], etc. It is not difficult to see that the study of such models and tasks is beneficial and actual.

Negproblem. We have introduced the concept of *negative information* in the moving and damageable dynamic system under certain conditions; of the *negative index of damageability* responsible for a definite relationship between the processes of hardening–softening of system elements under appropriate conditions; of *negative safety* of the system when the risk of its existence is higher than its critical value. In classical thermodynamics, the concept of *negative entropy* is well-known. Here, we can call the negativity of *Poisson's* coefficient in mechanics. Further, we can talk about supercritical (transmitting) states of the MTD system when *negative information*, (5.74)–(5.75), is revealed.

Let us assume the following hypothesis on the interrelation of damages and information. For example, let a solid Tribo-Fatigue system possess the strength (carrying capacity) equal to u_0, if it is assessed in units of potential energy density. It is not difficult to understand that in other words, u_0 is the maximum effective energy spent for creating a given system. Now, according to (3.39) and (5.34), the *generalized expressions* can be written for the *damageability increment*

$$\Delta\psi = \pm b_\psi \left[u_\Sigma^{eff}(\Lambda) - u_0 \right] \tag{6.1}$$

and for the *expenditure of stored (in the course of creating a system) information*

$$\Delta I = \mp b_I \left[u_\Sigma^{eff}(\Lambda) - u_0 \right] = \mp\, b_I u_0 \left(\psi_u^{eff} - 1 \right) = \mp\, b_I u_0 (-\delta) \tag{6.2}$$

during tests or operation that are characterized by the conditions $u_\Sigma^{eff} = u_\Sigma^{eff}\left(u_{\sigma(ch)}^{eff}, u_{\tau(ch)}^{eff}, u_{T(ch)}^{eff}, \Lambda_{k\backslash l\backslash q} \right)$ according to (2.2). Here, b_ψ and b_I are the transition functions, δ is the continuum parameter (Table 2.5). Use the state classification of damages according to Table 2.4, however, the numerical values of damageability characteristics will be different according to (6.1).

Let us analyze "processes" (6.1)–(6.2) for the case when the "minus" sign is kept in these relations. As shown in Fig. 6.2, they can be easily presented graphically, requiring that $b_\psi \neq b_I$ and considering that $\Lambda \gtreqless 1$. The curves describe a *transition of a given system from the positive state into the negative state simultaneously with respect to the both characteristics* ($\Delta\psi$ and ΔI). In this case, each of the three types of the curves (solid, dotted, and dash–dotted lines) corresponds to one (determined) value of the function of Λ-interactions. According to (6.1)–(6.2), the characteristic values of damageability and information will be

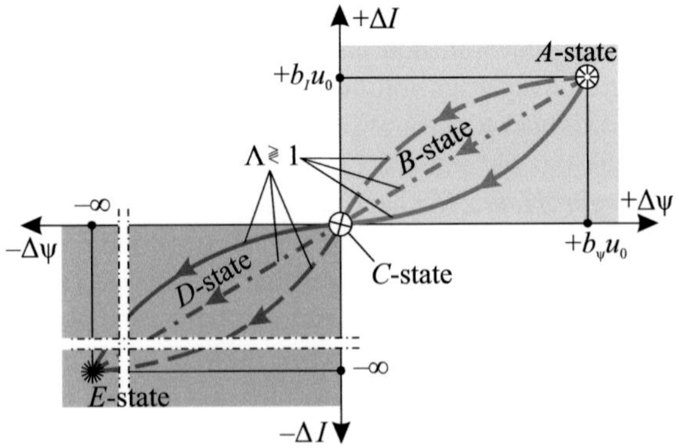

Fig. 6.2 Scheme of a possible transition of the Tribo-Fatigue system into the negative state (supercritical *D*-states)

$$
\begin{aligned}
&A-\text{state:} \quad u_{\Sigma}^{eff}=0, \quad \Delta\psi_A=+b_{\psi}u_0>0,\ \Delta I_A=+b_I u_0>0,\\
&C-\text{state:} \quad u_{\Sigma}^{eff}=u_0, \quad \Delta\psi_C=0=\Delta I_C,\\
&D-\text{state:} \quad u_{\Sigma}^{eff}>u_0, \quad \Delta\psi_D^*<0,\ \Delta I_D^*<0,\\
&E-\text{state:} \quad u_{\Sigma}^{eff}\to\infty, \quad \Delta\psi_E^*,\ \Delta I_E^*\to-\infty.
\end{aligned}
$$

It should be noted that according to (6.1)–(6.2), positive and negative states "mixed" with respect to two characteristics ($\Delta\psi$ and ΔI) are possible. Such states appear in the case when different signs are taken for (6.1)–(6.2). The curves for the system states will then be located in the II and the IV quadrants in Fig. 6.3.

$$
\begin{aligned}
&A-\text{state:} \quad u_{\Sigma}^{eff}=0, \quad \Delta\psi_A=-b_{\psi}u_0<0, \quad \Delta I_A=+b_I u_0>0,\\
&C-\text{state:} \quad u_{\Sigma}^{eff}=u_0, \quad \Delta\psi_C=+0, \quad \Delta I_C=-0,\\
&D-\text{state:} \quad u_{\Sigma}^{eff}>u_0, \quad \Delta\psi_D^*>0, \quad \Delta I_D^*<0,\\
&E-\text{state:} \quad u_{\Sigma}^{eff}\to\infty, \quad \Delta\psi_E^*\to+\infty, \quad \Delta I_E^*\to-\infty.
\end{aligned}
$$

We have to admit that now preference can be given to one of the schemes in Figs. 6.2 or 6.3 depending on the conditions of existence of this or that system. Of course, this question should be studied further. It should be added that such schemes can also be constructed in terms of the continuum parameter δ (6.2) instead of the analysis of damageability ($\Delta\psi$).

In our opinion, studies and their results generalized in the above directions yield not only the general concept of *negative states of systems*, but also, if we say more boldly, they lead to negative mechanics (?) and, even possibly, to the *problem of negativity in the Universe*?

There is one more hard problem: how are *transitions from the "physical negativity" to the "physical positivity" and*, vice versa, caused and realized? What are the limitations of such transitions? Our monograph also considers this problem as

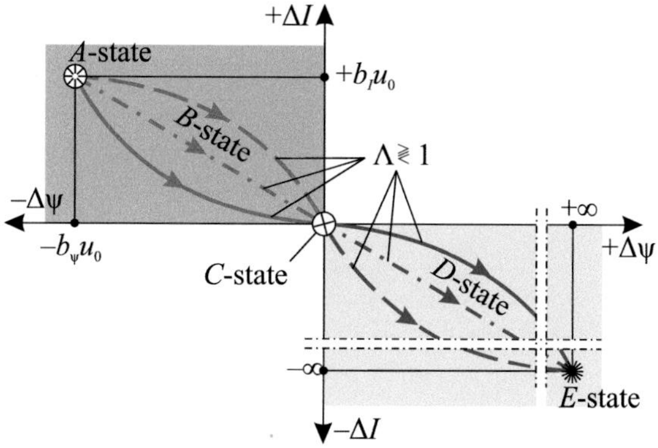

Fig. 6.3 Scheme of another possible transition of the Tribo-Fatigue system in the supercritical *D*-states

applied to MTD systems. In general, do such transitions define the *hysteresis of evolution*?

In the analysis of the mentioned problems, it is required to distinguish "*irreversible physical states of certain direction*" and *simply positive and negative quantities that are usually introduced in effort to conveniently describe these or those reversible events*.

Dialectics of state. In conclusion, it should be said that it is now actual and necessary to comprehend the *states of the MTD system and its evolution from the viewpoint of philosophy*. As known, the interaction and communication of philosophy, engineering and natural sciences was always and remains fruitful [24, 27].

First of all, the basic concepts such as interaction and mutual influence, risk and safety, damageability, entropy and information, negative phenomena (or negative states), etc., should be more deeply understood from the viewpoint of philosophy. Finally, the philosophic analysis of possible states of systems caused and described by numerous and diverse parameters should be made. For example, it is difficult to answer: can "positive" phenomena (situations) become "negative" passing the indifferent (zero) state? Vice versa: can (must) "negative" phenomena (situations) go into "positive" passing the same zero? Are there fundamental prohibitions? It means that the zero is a meaningful quantity, for example, for indifferent states (matter? objects? systems?). The concept of zero needs an additional philosophical (and physical) understanding.

On the basis of the generalized theory of A-evolution a deeper synthesis of natural-science and philosophy knowledge can be made using the modern *technology* of *NBIC-convergence* (for example, [5]). Figure 6.4 presents the sketch of the problem of Nano (N), Bio (B), Inform (I), Cogno (C) that is based on a

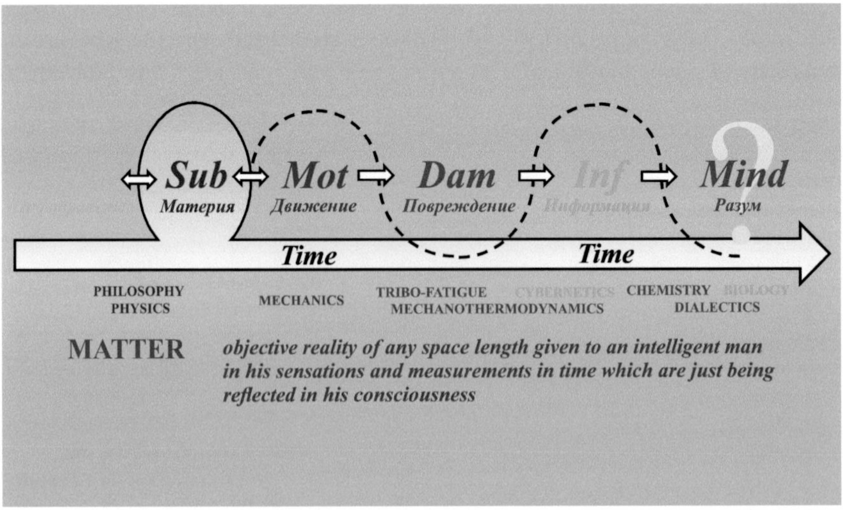

Fig. 6.4 Fundamental problem

somewhat *specified known philosophical concept of matter;* wherein information is defined as a change in internal damageability *(or irreversible states) of the system*, and *the elements of intelligence are interpreted as a result of qualitative jumps in diverse processes of (qualitative) accumulation of information.*

Of course, we have formulated only some of the areas of further research. It is easy to see that each of them is multidimensional. Bearing in mind this monograph, we would like to invite specialists (especially young) to dialog. One cannot ignore that the development of fundamental problems of mechanothermodynamics promises… what it promises—let every thoughtful reader assess: the new direction in science is undoubtedly in his "hands".

We, the authors of this monograph, practice a simple thing that new knowledge elucidates space and time.

References

1. Chigarev, A.V., Mikhasev, G.I., Borisov, A.V.: Syomechanica. Grevtsov Publishing House, Minsk (2010)
2. Hele-Show, H.S.: Human endurance curves. Proc. Inst. Mech. Eng., October 1911, Parts. III and IV (1911)
3. Makhutov, N.A., Gadenin, M.M.: Fundamental and applied principles of a comprehensive risk analysis and man-made protection. In: Zhuravkov, M.A., et al. (eds.) Proceedings of 6th International Symposium on Tribo-Fatigue, 25 October–1 November 2010, Part 1, pp. 103–120. BSU Press, Minsk (2010)
4. Makhutov, N.A.: Strength and Safety: Fundamental and Applied Studies. Nauka, Novosibirsk (2008)
5. Roco, M.: Converging technologies for improving human performance. In: Roco, M., Bainbridge, W. (eds.) Nanotechnology, Biotechnology, Information Technology and Cognitive Science. Arlington (2004)
6. Sherbakov, S.S., Sosnovskiy, L.A.: Mechanics of Tribo-Fatigue Systems. BSU Press, Minsk (2010). (in Russian)
7. Sherbakov, S.S.: To the general theory of interaction in Tribo-Fatigue systems. Harmonic development of systems is the third way of mankind. In: Proceedings of 1st International Congress, 8–10 October 2011, pp. 158–163. Institute of Creative Technologies, Odessa (2011)
8. Sosnovskiy, L.A., et al.: Reliability, Risk, Quality. BelSUT Press, Gomel (2012). (in Russian)
9. Sosnovskiy, L.A., Scherbakov, S.S.: Non-classical model of a deformable solid and the scattered wear-fatigue damage criterion. Fatigue and thermal fatigue of materials and construction elements. In: International Scientific Technical Conference, 28–31 May 2013, Kiev, pp. 276–278. Press of the G.S. Pisarenko IPP NAS Ukraine, Kiev (2013)
10. Sosnovskiy, L.A., Sherbakov, S.S.: Mechanothermodynamic system and its behavior. Continuum. Mech. Thermodyn. **24**(3), 239–256 (2012)
11. Sosnovskiy, L.A., Sherbakov, S.S.: Possibility to Construct Mechanothermodynamics. Nauka i Innovatsii **60**, 24–29 (2008). (in Russian)
12. Sosnovskiy, L.A., Sherbakov, S.S.: Surprises of Tribo-Fatigue. Magic Book, Minsk (2009)
13. Sosnovskiy, L.A.: L-Risk (Mechanothermodynamics of Irreversible Damages). BelSUT Press, Gomel (2004). (in Russian)

14. Sosnovskiy, L.A., Makhutov, N.A.: Reliability of Tribo-Fatigue systems in the non-classical statement. In: Conference on Survivability and Construction Material Science, 22–24 October 2012, Moscow, p. 53. Press of the A.A. Blagonravov Institute of Mechanical Engineering, Moscow (2012)
15. Sosnovskiy, L.A., Matukhov, N.A.: Tribo-Fatigue and wear-fatigue damages in the problem of machine resources and safety. Press of FTsNTP "Safety"—SPA "TRIBO-FATIGUE", Moscow–Gomel (2000)
16. Sosnovskiy, L.A., Sherbakov, S.S.: Surprises of Tribo-Fatigue. BelSUT Press, Gomel (2005). (in Russian)
17. Sosnovskiy, L.A., Sherbakov, S.S.: Dynamic problem in the nonclassical statement of variable thickness. In: 11th Belarusian Mathematical Conference, 4–9 November 2012, Part 3, pp. 91–92. Press of the Institute of Mathematics of NAS, Minsk (2012)
18. Sosnovskiy, L.A.: Intelligent Dynamic Systems: Problem and Research Perspectives. Mechanics 2011. In: Proceedings of 5th Belarusian Congress on Theoretical and Applied Mechanics, vol. 1, pp. 64–79. Press of the Joint Institute of Mechanical Engineering of NAS of Belarus, Minsk (2011) (in Russian)
19. Sosnovskiy, L.A.: Life field and golden proportions. Nauka i Innovatsii **79**(9):26–33 (2009) (in Russian)
20. Sosnovskiy, L.A.: Mechanics of Wear-Fatigue Damage. BelSUT Press, Gomel (2007). (in Russian)
21. Sosnovskiy, L.A.: On one method of construction a relationship between motion, information and damages in the mechanothermodynamic system. Mech Mach Mech Mater **17**(4), 87–90 (2011)
22. Sosnovskiy, L.A.: On one type of entropy as the absorption measure of energy spent for damage generation in the mechanothermodynamic system. Doklady NAS Belarus **51**(6), 100–104 (2007)
23. Sosnovskiy, L.A.: The field of the fate: first representation. Nauka i Innovatsii **80**(10), 29–33 (2009)
24. Sosnovskyi, L.A., et al.: On the interaction of sciences. Harmonic development of systems is the third way of mankind. In: Proceedings of I st International Congress, 8–10 October 2011, pp. 115–126. Institute of Creative Technologies, Odessa (2011)
25. Troshchenko, V.T.: Fatigue and inelasticity of metals. Naukova dumka, Kiev (1971)
26. Tsaturyan, A.K., Shtein, A.A. (eds.): Biomechanics: Achievements and Perspectives. Modern Problems of Biomechanics. Issue 11. Moscow State University Press, Moscow (2006)
27. Vladimirov, V.A., et al.: Risk Control: Risk, Stable Development, Synergetics. Nauka, Moscow (2000)

About the Authors

Sosnovskiy Leonid Adamovich

Published more than 1000 scientific works and inventions including 39 monographs, reference books, and handbooks.

Professor, Doctor of Engineering Sciences, Ph.D., Director of S&P Group TRIBOFATIGUE Ltd, Scientific Head of Interdepartmental Laboratory «TRIBO-FATIGUE», Professor of Belarusian University of Transport, Honored Scientist of the Republic of Belarus, Laureate of the State Prize of Ukraine, Member of Russian National Committee on Theoretical and Applied Mechanics, Co-chairman of the International Coordinate Council on Tribo-Fatigue.

Scientific interests: Mechanothermodynamics, Tribo-Fatigue, Dialectics.

Honorary Citizen of the Town of Chechersk.

e-mail: sosnovskiy@tribo-fatigue.com

Sherbakov Sergei Sergeevich

Published more than 230 scientific works, including 6 monographs, 3 inventions, course of lectures and two handbooks.

Doctor of Physical and Mathematical Sciences, Ph.D., Associate Professor of Department of Theoretical and Applied Mechanics of Belarusian State University, Secretary of the Council of the Faculty of Mechanics and Mathematics, Laureate of A.N. Sevchenko Prize in Natural Sciences for Young Scientists, Laureate of President of Republic of Belarus Scholarship for Young Scientists.

Scientific interests: Mechanothermodynamics, mechanical and mathematical modeling in Tribo-Fatigue, Mechanics of Deformable Solid Body, Tribology, Mechanics of Fatigue Damage and Fracture.

e-mail: sherbakovss@mail.ru

L. Sosnovskiy and S. Sherbakov, *Mechanothermodynamics*,
DOI 10.1007/978-3-319-24981-0

Index

L. Sosnovskiy and S. Sherbakov, *Mechanothermodynamics*,
DOI 10.1007/978-3-319-24981-0

Printed in the United States
By Bookmasters